TECHNOLOGY TRANSFER of PLANT BIOTECHNOLOGY

A CRC Series in

Current Topics in Plant Molecular Biology

Peter M. Gresshoff, Editor

TECHNOLOGY TRANSFER OF PLANT BIOTECHNOLOGY, 1996

Previous Titles in Series

PLANT BIOTECHNOLOGY AND DEVELOPMENT, 1992

PLANT RESPONSES TO THE ENVIRONMENT, 1993

BIOTECHNOLOGICAL APPLICATIONS OF PLANT CULTURES, 1994

TECHNOLOGY TRANSFER of PLANT BIOTECHNOLOGY

Edited by
Peter M. Gresshoff
Plant Molecular Genetics
Institute of Agriculture
Center for Legume Research
The University of Tennessee
Knoxville, Tennessee

CRC Press
Taylor & Francis Group
Boca Raton   London   New York

CRC Press is an imprint of the
Taylor & Francis Group, an **informa** business

CRC Press
Taylor & Francis Group
6000 Broken Sound Parkway NW, Suite 300
Boca Raton, FL 3487-2742

First issued in paperback 2020

ISBN-13: 978-0-367-57951-7 (pbk)
ISBN-13: 978-0-8493-8265-9 (hbk)

This book contains information obtained from authentic and highly regarded sources. Reasonable efforts have been made to publish reliable data and information, but the author and publisher cannot assume responsibility for the validity of all materials or the consequences of their use. The authors and publishers have attempted to trace the copyright holders of all material reproduced in this publication and apologize to copyright holders if permission to publish in this form has not been obtained. If any copyright material has not been acknowledged please write and let us know so we may rectify in any future reprint.

Visit the Taylor & Francis Web site at
http://www.taylorandfrancis.com

and the CRC Press Web site at
http://www.crcpress.com

Library of Congress Cataloging-in-Publication Data

Technology transfer of plant biotechnology / editor, Peter M. Gresshoff
        p. cm. -- (A CRC series of current topics in plant molecular biology)
    Includes bibliographical references and index.
    ISBN 0-8493-8265-3
    1. Plant biotechnology. I. Gresshoff, Peter M., 1948– . II. Series.
SB106.B56T43   1996
   631.5'233--dc621   96-48232
                                                             CIP

Library of Congress Card Number 96-48232

Cover design: Denise Craig

Table of Contents

Preface

Chapter 1   1

Technology Transfer: Biotechnology Information and the Federal Government

Raymond C. Dobert

Chapter 2   13

From Lab Bench to Marketplace: The Calgene FLAVR SAVR™ Tomato

Belinda Martineau

Chapter 3   25

Biotechnological Applications of Inheritable and Inducible Resistance to Diseases in Plants

Sadik Tuzun, Peter A. Gay, Christopher B. Lawrence, Tracy L. Robertson and Ronald J. Sayler

Chapter 4   41

The Role of Antifungal Metabolites in Biological Control of Plant Disease

Steven Hill, Philip E. Hammer and James Ligon

Chapter 5   51

Genetically Engineered Protection of Plants against Potyviruses

Indu B. Maiti and Arthur B. Hunt

Chapter 6   67

Negative Selection Markers for Plants

Mihály Czakó, Allan R. Wenck and László Márton

Chapter 7   95

Considerations for Development and Commercialization of Plant Cell Processes and Products

Walter E. Goldstein

Chapter 8   111

DNA Diagnostics in Horticulture

Wm. Vance Baird, Albert G. Abbott, Robert Ballard, Bryon Sosinski and Sriyani Rajapakse

Chapter 9   131

Commercial Applications of DNA Profiling by Amplification with
Arbitrary Oligonucleotide Primers

Peter M. Gresshoff and Gustavo Caetano-Anollés

Chapter 10   149

Phylogenetic Relationships in the Tribe *Triticeae*

Ponaka V. Reddy and Khairy M. Soliman

Chapter 11   167

Isolation of Plant Peptide Transporter Genes from *Arabidopsis*
by Yeast Complementation

Henry-York Steiner, Wei Song, Larry Zhang, Fred Naider, Jeffrey Becker and Gary Stacey

Chapter 12   177

Confocal Laser Scanning Light Microscopy with Optical Sectioning:
Applications in Plant Science Research

Sukumar Saha, Anitha Kakani, Val Sapra, Allan Zipf and Govind C. Sharma

Chapter 13   191

Field Testing of Genetically Engineered Crops: Public-private
Institution Comparisons

Patrick A. Stewart and A. Ann Sorensen

Glossary   207

Index   223

The Editor

Peter Michael Gresshoff, Ph.D., D.Sc., holds the endowed Racheff Chair of Plant Molecular Genetics at the University of Tennessee in Knoxville.

Professor Gresshoff, a native of Berlin (Germany), graduated in genetics/biochemistry from the University of Alberta in Edmonton in 1970 and then undertook postgraduate studies at the Australian National University in Canberra, Australia, where he obtained his Ph.D. in 1973, and his Doctorate of Science in 1989. He was elected as a Foreign Member of the Russian Academy of Agricultural Sciences (1995) as well as a Fellow of the American Association for the Advancement of Science (AAAS)

He completed his postdoctoral work as an Alexander von Humboldt Fellow (1973 to 1975) at the University of Hohenheim (Germany) and Research Fellow (1975 to 1979) in the Genetics Department (RSBS, ANU, Canberra), then headed by late Professor William Hayes. He was appointed Senior Lecturer of Genetics in the Botany Department at the Australian National University in 1979, where he built up an international research group investigating the genetics of symbiotic nitrogen fixation. He assumed his present position in Knoxville in January 1988, continuing his research direction by focusing on the macro- and micromolecular changes involved in nodulation. His recent interests have turned to plant genome analysis and DNA fingerprinting.

Professor Gresshoff chaired the 8th International Congress on Nitrogen Fixation in Knoxville, and co-chaired the 8th International Congress of Molecular Plant Microbe Interactions (Knoxville, July 1996). He also organized the Annual Gatlinburg Symposia, which focus on current topics in plant biology. He is the assistant director of the Center for Legume Research at UT.

He was awarded the Alexander von Humboldt Fellowship twice (1973 and 1985) and is a member of the editorial boards of *Physiologia Plantarum, Symbiosis* and the *Journal of Plant Physiology*. He has received major research funding from competitive grants, the United Soybean Board, the Tennessee Soybean Association, the Human Frontier Science Program, biotechnology firms, and the Australian Government. He has published over 225 refereed publications, several book chapters as well as 7 books. His work on DNA fingerprinting and nitrate tolerant soybeans is covered by US patents. He is a member of Phi Kappa Phi, Gamma Sigma Delta and Sigma Xi and was awarded the Outstanding Scholar Award for the Year by Phi Kappa Phi. As a dedicated teacher and researcher, he believes in technology transfer and innovative science. He has been an advisor on plant biotechnology to the European Commission, the Japanese Government, and the Environmental Protection Agency as well as the Department of Energy in the United States.

Preface

Plant biotechnology has grown to be a vast field of research and developmental science. It comprises more than recombinant DNA technology, although most of it is based on our ability to isolate, modify and transfer genetic molecules. Figure 1 gives a broad definition of biotechnology, while Figure 2 delineates the four major areas as dealt with in this book.

The technology field is about 20 years old. In the mid-1970s the regeneration of plants from isolated cells or embryos became more commonplace. Although some of the major crop species, like maize and soybean, were 'recalcitrant', successful gene transfer into crops was possible by the late 1980s. Transformation of genes using either *Agrobacterium tumefaciens*, electroporation, or microprojectile bombardment, at first focused on single dominant traits introduced from external genetic sources. Examples include virus, herbicide, and insect resistances; later genetic transfers were used to modify quality related traits, such as the shelf-life of a fruit, bio-degradable plastic biosynthesis, vegetable oil quality, or flower color.

Biotechnology is:
the directed application of
technologies to select or develop
novel heritable genetic properties
in organisms or
their cellular processes to
generate products for the
benefit of humankind and the planet

Figure 1: A broad definition of biotechnology. More restrictive terms could include the terms genetic engineering, recombinant DNA technology and "not normally found in Nature".

Plant biotechnology expanded in the field of cell and tissue culture. Small culture vessels gave way to reaction fermenters. Focus continued on the biotransformations and synthesis of secondary metabolites of use in pharmaceutical applications. Major obstacles seem to remain here as the maintenance costs of cultures and percentage yields seldomly reach the break-even point.

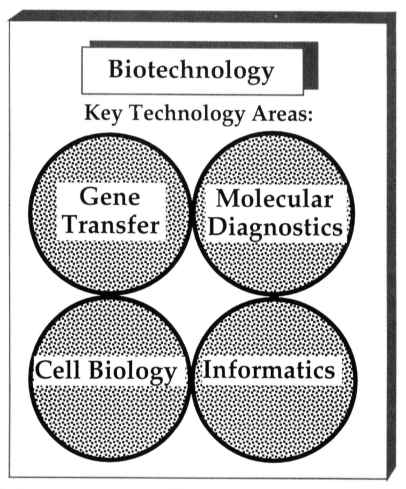

Figure 2: Key component areas of biotechnology as discussed in this book. Informatics deals with the immense computer and information-based revolution occurring presently, which allows the storage and comparison of large data sets.

Outstanding advances have been obtained through genome analysis. The tools of molecular biology and genetics have now permitted the generation and detection of a large variety of molecular markers across the genomes of many plants. This was the subject of the previous volume in this series. Such markers are used for marker-based selection in plant breeding programs allowing backcross conversion at more efficient speeds. Molecular markers function as sign posts for the isolation of genes, which hitherto have only been known by their phenotype.

DNA profiling also finds application in identity testing. While this is popularized through high visibility criminal trials, a more mundane application is found in plant variety testing. Such application not only is useful in field crops, but progressively has found uses in horticultural crops and natural populations. For example, the European Commission supports DNA profiling work with whales, and researchers in Costa Rica are evaluating genetic diversity of tropical rain forests at Villa Caletas.

Technology transfer is usually defined as the conversion of laboratory skills to the marketplace. Universities and industrial firms have special units to facilitate technology transfer. National science programs are compared and judged by their ability to transfer technology. However, there are other needs of technology transfer. Many of the advances described in this book seem only to benefit the developed world. There is a need to transfer technology to the developing world. It is these regions that have food deficiencies and extreme environmental problems because of groundwater pollution stemming from agricultural production. Yet these regions of the world cannot afford most of the infrastructure needed for plant biotechnology. Shipments of quality chemicals are delayed, electric supply is uncertain, foreign currency is missing, and capable scientists from these regions are drained to laboratories of the developed world. Thus technology transfer to those regions becomes a major problem, hopefully solved through collaboration and work sharing. Many processes in biotechnology are based on basic biological and genetic facts which can be obtained in Third World laboratories and field stations. Biological diversity probably is maximal in tropical countries; this can be harnessed to yield new genes and gene combinations.

This book includes an overview of new technology as it is pivoted on the edge of commercialization. The recent past saw the first commercial releases and marketing of a genetically engineered crop in the US. Tomatoes exist with better shelf-life, cotton and corn grow with insect resistance and transgenic soybean possess herbicide resistance. Others are following rapidly. Field tests are conducted around the world including Australia, Chile and South Africa. DNA profiling of plant material has been used in criminal investigations and is used to certify plant material prior to shipping. The 'public' is becoming increasingly literate of terms like 'gene', 'DNA' and 'biotechnology'. Perhaps the fear that biotechnology will produce Frankenstein-like monsters is mellowed by our efforts to demonstrate that this technology, while being monitored, holds great promises for the benefit of this planet.

I thank members of my staff in the Racheff Chair and participants of the Fifth Gatlinburg Symposium. Cathy Balogh is thanked for technical and production assistance. Janice Crockett, my secretary, was essential with secretarial and administrative help. Paul Petralia was tolerant and supportive during editing.

Peter M. Gresshoff   July 1996
Knoxville, Tennessee

Technology Transfer: Biotechnology Information and the Federal Government

Raymond C. Dobert

Biotechnology Information Center, National Agricultural Library,
Agricultural Research Service - United States Department of Agriculture (USDA), Beltsville, MD 20705,
USA

Overview

Technology transfer is a process that requires effective communication and large amounts of information. The federal government, as the largest supporter of scientific research in this country, has a vested interest in promoting the technology transfer process. To accomplish this goal, a number of information services and dissemination tools have been developed that can be used to facilitate the technology transfer process. These same resources can be tapped into by researchers in industry and academia to obtain valuable research information. Since information is the linchpin of successful technology transfer, the USDA's National Agricultural Library houses specialized information centers that focus on biotechnology, plant genomics and technology transfer.

Several databases are also maintained by USDA to provide access to current and pre-publication research findings from USDA-supported laboratories. Many government and institutional resources are now accessible through the Internet, a venue which is likely to serve as the foundation for future federal technology transfer activities. Understanding and participating in technology transfer activities will become increasingly important as these activities are likely to become a prerequisite for some federally-funded research grants in the future.

What is 'technology transfer'?

Technology transfer means different things to different people. To some scientists it may be little more than a buzzword, to others it may be the underlying goal of their research program. With publicly-supported research under increasing pressure to generate or support commercially relevant technologies, technology transfer will become a necessary part of any research program. Technology transfer can also have its financial rewards, as many institutions provide for researchers to share in revenue generated through licensing fees and royalties.

Simply stated technology transfer is the transfer of research results from research institutions, such as universities or federal laboratories, to the commercial sector or other technology users. For the agricultural sciences, the ultimate technology user may be the farmer or consumer. To be effective, technology transfer relies on **effective communication** and vast amounts of **information**. But at times the sheer volume of information available can hinder its effective use. Like society in general, researchers can often be overloaded with information. Some 1 billion pages of "new" information is created per day in the U.S.; 736,000 book titles are published per year! At this rate, published scholarly information doubles every 8 years. Unfortunately knowledge and information are not synonymous terms, since research results have value as knowledge only when they are put to effective use. Transferring research results into practical applications is what technology transfer is all about.

The federal government supports technology transfer by providing both information and communication channels and reducing the impediments that can limit technology adoption by the private sector. Ideally federal agencies should serve as brokers of scientific, technical and business information. Development of new technologies is also supported through in-house (intramural) research projects which are directly supported by the federal government. In addition to the federal efforts to move technology from the lab to the marketplace, many universities and other research institutions have their own technology transfer offices to assist researchers at a local level.

Much of technology transfer relies upon the creation, protection and subsequent dissemination and application of intellectual property. The primary means to protect intellectual property is the patent, a form of protection allowed for genetically modified plants only since 1985.[1] Thus, a researcher first formulates an idea, which is developed into a practical finding or process. If this discovery can be demonstrated to be useful, novel, and nonobvious it stands a chance of being patentable. Once a technology is patented, its practical use is usually achieved through licensing agreements. Unfortunately, prior to 1980 intellectual property resulting from federally-supported research projects (both intramural and extramural) was essentially owned by the federal government. Given this restriction, researchers at universities and federal labs had little incentive to pursue the development and commercialization of their work.

In the intervening 15 years a number of federal technology transfer laws, beginning with the landmark Bayh-Dole Technology Transfer Act of 1980, have

---

[1] *Ex parte* Hibberd, 227 U.S.P.Q. 443 (P.T.O. Bd.. App. & Inter. 1985)

made it increasingly easier for private companies to access and benefit from federal research projects. First and foremost, federal research can now be patented and made available for licensing while protecting a researchers' rights to his or her invention. As an example, cooperative research and development agreements (CRADAs) between a private firm and a federal research agency now provide the first right to exclusive licenses on patented inventions to the cooperating company. In return, the cooperating enterprise can provide the know-how needed to proceed through the development and commercialization phases required to bring a new product or service to the market. In the agricultural research area, the USDA's Agricultural Research Service (ARS) currently has over 500 researcher-initiated CRADAs in place to help develop promising technologies. However, the ability to provide intellectual property protection for the wealth of federally-funded research does not ensure that these technologies are utilized. The award of a patent is not the end of the road for the technology transfer process. The technology must be linked to a user willing to proceed with the commercialize process. How do commercial firms learn about available government technology? Timely research information is the key.

The government as information provider

Several factors make the federal government a logical source of research information. The federal government is the largest funding source for basic research. In 1993 over US$4.2 billion was spent to support "biotechnology" research, although only a small portion of that money ($233 million) supported research in the agricultural field.[2] This fact coupled with increasing calls for spending accountability, requires that the "research portfolio" of the federal government be made as accessible as possible. Thus, many branches of the federal government work to facilitate information sharing and rapid communication of research results with the private sector.

Federal agencies also hold a significant portfolio of intellectual property in the area of biotechnology. A recent technology directory found over 2,100 biotechnology-related patents which were held by the US government.[3] Many times the information or research results contained in these patents is unpublished and is not available from any other source. Unlike patents held by individuals or corporations, which are available to the public only after the patent has been awarded, pending patent applications from federal research agencies are often available for review.

---

[2] Federal Coordinating Council for Science, Engineering, and Technology. *Biotechnology for the 21st Century: Realizing the Promise - A Report by the Committee on Life Sciences and Health of the Federal Coordinating Council for Science, Engineering, and Technology*, Executive Office of the President, Washington, DC, June 1993.

[3] Rader, R.A. & S.A. Young. *Federal Bio-Technology Transfer Directory*, Biotechnology Information Institute, Rockville, MD, 1994.

As previously alluded, only a small percentage of the federal dollars spent in the field of biotechnology are directed towards agriculture or the plant sciences. As a result of this relatively small funding base, plant biotechnology information is generally obtained from sources that cover the much broader field of agriculture or focus on biotechnology, which, in general, emphasize medical/pharmaceutical applications. Examples of federally supported biotechnology information resources which cover plant molecular biology only incidentally would include GENBANK and the products and services of the NIH's National Center for Biotechnology Information (NCBI). These services serve as a primary research tool and have not been widely used in technology transfer activities. Nonetheless, a number of information products and services available on plant biotechnology from the federal government and other institutions are available.

USDA maintains several databases that profile current research including the CRIS (Current Research Information System), and TEKTRAN databases. In addition, USDA has several offices dedicated to technology transfer activities and funding private sector research through its SBIR (Small Business Innovation Research) grants program. The Biotechnology Information Center and the Technology Transfer Information Center at the National Agricultural Library provide assistance in accessing these and other information sources. An additional source of information, the FEDRIP (Federal Research in Progress) database contains over 120,000 research projects supported by federal agencies. Many other government programs, both federal and state, are focused on facilitating the transfer of information out of laboratories. For the committed researcher or technology developer, assistance and opportunities for technology transfer are plentiful.

The Biotechnology Information Center (BIC)

The Biotechnology Information Center (BIC) in Beltsville, Maryland was founded in 1985 to help complement and support the biotechnology research and outreach programs at the USDA.[4] Housed at the National Agricultural Library, site of the world's largest collection of agricultural literature, the BIC has access to an extensive collection of biotechnology information. The Center is designed to help answer the who, what, when and where questions regarding agricultural biotechnology research and federal policy. To accomplish this task, the BIC provides the following services to the biotechnology research community:

I. Provision of Informational Products - Aids such as directories of experts, institutions, associations, upcoming meetings and bibliographic material are produced to assist consumers, educators, public and private-sector researchers, and policy-makers find current information on the field of biotechnology.

---

[4] The Biotechnology Information Center can be reached at 1-301-504-5947 OR via e-mail to biotech@nalusda.gov.

Many of the informational products are also available electronically via the Internet (see below).

II. Specialized References Services - To handle specialized information requests, the BIC staff are familiar with concepts and techniques used in biotechnology. They can provide assistance in locating biotechnology information for business, research and study. BIC will perform brief complimentary literature searches of USDA's Agricola database on specific biotechnology topics or conduct exhaustive searches of major databases on a cost recovery basis. From the novice to the expert, the staff of the BIC can provide the appropriate level of information.

III. Biotechnology Information Center Gopher/WWW
To support the transition of the National Agricultural Library from a resource based on paper to one which is available electronically from remote sites, the Biotechnology Information Center has focused its energy on developing Internet-available resources.[5] In cooperation with the University of Maryland Computer Science Center, the BIC has constructed WWW/gopher sites which increase the accessibility to BIC information products such as bibliographies and resource guides. As Internet tools, gopher and the world-wide-web are simply applications that help organize and retrieve hierarchical or hyperlinked files from various "servers" connected to the Internet. It provides a uniform interface that allows for universal access to text files, software programs, databases and for the WWW graphical and audio electronic files. In addition to BIC publications, several biotechnology newsletters are archived at the site and connections to other biotechnology-related web and gopher sites around the world have been made. Federal documents related to agricultural biotechnology (reports, press releases, regulatory documents) and a gateway to full-text biotechnology patents are also included.

The Technology Transfer Information Center (TTIC)

The National Agricultural Library also houses the Technology Transfer Information Center[6] which serves to promote the rapid conversion of federally-developed inventions into commercial products by getting the results of research into the hands of individuals and organizations who can put it into practical use. To accomplish these goals the Center provides a variety of services to professionals involved in the innovation process. These include:

---

5 The Biotechnology Information Center WWW is available at: http://WWW.nal.usda.gov/bic
   OR use the commands: *gopher* gopher.nalusda.gov; select NAL Information Centers/BIC
   OR *gopher* inform.umd.edu OR *telnet* inform.umd.edu; select Educational Resources/
   Academic Resources By Topic/Agriculture Environment Resources/.

6 The Technology Transfer Information Center can be reached at 1-301-504-6875 or via e-mail to
   ttic@nalusda.gov OR access their web site at: http://WWW.nal.usda.gov/ttic

I. Reference services to locate information relative to individual technology transfer needs through the use of database searches and informal information collection from individuals, Federal agencies, or public and private technology information organizations.

II. Collection development to acquire the latest information of developments in technology transfer, research and development and federal activities related to technology developments.

III. Facilitating information exchange by bringing together people and information necessary to promote innovative solutions to problems. The TTIC supports research efforts, participates in laboratory research consortia, and coordinates the development and exchange of technology information throughout the scientific, business, governmental, and academic communities. The TTIC works closely with the Federal Laboratory Consortium (FLC), an organization representing over 600 member research laboratories that works to bring the labs together with potential users of government-developed technologies.

Other USDA resources

A number of additional biotechnology information resources are provided by the USDA to facilitate the exchange of research information.

National Biological Impact Assessment Program (NBIAP)

The National Biological Impact Assessment Program (NBIAP)[7], sponsored by the Cooperative State Research Education and Extension Service (CSREES) has several programs designed to promote biosafety by expediting the flow of biotechnology information. The **NBIAP Information System or Bulletin Board** serves as the primary source of this biosafety and biotechnology information. The Information System is now accessible through a number of means including:

- **VIA WWW:**   http://gophisb.biochem.vt.edu
- **VIA TELNET:** *telnet* nbiap.biochem.vt.edu
- **VIA GOPHER:** *gopher* ftp.nbiap.vt.edu

Modem and telnet access provide the complete range of services and files available, while the gopher-accessible version is more limited. The System combines a monthly News Report with a number of searchable databases and text files of biotechnology information as well as an interactive bulletin board "forum" for interactive discussions.

---

[7] The NBIAP offices can be reached at 1-202-401-4892; FAX 1-202-401-4888 to obtain more information about the NBIAP programs and services.

Databases include information on regulatory contacts and guidelines, biotechnology companies, international researchers and federally approved field tests. This last database, called the Biomonitoring Database, includes information, including some full text environmental assessments (EA), provided by USDA's regulatory branch of the Animal Plant Health Inspection Service (APHIS) (see below). The first release of this CD-ROM contains the full text of 176 environmental assessments (EAs) issued by APHIS and allows for searches by recipient organism, donor organism, transformation method or state in which test occurred.

USDA/APHIS Biotechnology Permit Applications / Notifications

Prior to each field test of a genetically modified crop a permit application or notification must be filed with Biotechnology, Biologics and Environmental Protection (BBEP) Unit of USDA's Animal and Plant Health Inspection Service (APHIS). The information that is released by USDA includes the institution applying for the permit/notification, the parent plant (or microbe), source of introduced gene(s), and the general type of phenotypic change such as herbicide tolerance, insect resistance or product quality. This information is available directly from APHIS or can be accessed electronically via the APHIS/BBEP web site.[8] This information is often the only way of knowing what genetically engineered products or applications private institutions are developing and preparing to bring to the marketplace. Companies are able to protect their proprietary information by declaring some elements of an application to be confidential business information (CBI). Thus, the information gained from these documents varies depending on the amount of information protected as CBI.

Interested in knowing who is doing research on recombination in *Arabidopsis*? There are several current research databases that can be used to help answer the question. As with other resources already mentioned, more of these resources are becoming available via the Internet.

The **CRIS database** (Current Research Information System) is a compilation of USDA and state supported research in agriculture, food, forestry and related fields. The database contains some 30,000 project summaries of recently funded research projects conducted at federal labs, land-grant institutions or cooperating state institutions. The database is available from the on-line vendor Dialog® or in CD-ROM format. CRIS has recently become available through the Internet (gopher to gopher.sura.net then select Databases and Network Information).[9]

---

[8] Contact: Biotechnology Coordination and Tech. Assistance, USDA/APHIS/BBEP 4700 River Road, Unit 147, Riverdale MD 20737-1228, USA. Tel: 1-301-734-7601 Fax: 1-301-734-8669. Additional information is available at: http://WWW.aphis.usda.gov/BBEP

[9] The CRIS data base is at: *gopher* gopher.sura.net   select Databases and Network Information OR http://cristel.nal.usda.gov

Each entry describes the objectives and approach of the project as well as annual progress reports and publications resulting from the work (Figure 1). The Inventory of Canadian AGRI-Food Research (ICAR) database, containing some 4,000 Canadian research projects in agriculture, food, human nutrition and biotechnology is included together with CRIS on the gopher and CD-ROM versions.

An additional resource that provides information about USDA funded research is the National Research Initiative Competitive Grants Program **(NRICGP) Abstracts of Funded Research.** Previously available only in print form, this resource is now available via the Internet by gopher and WWW.[10] The abstracts available on this searchable database are nontechnical abstracts written by the principal investigator of each individual grant, starting with Fiscal Year 1993. Each entry also includes the title, principal investigators, awardee institution, dollar amount, and proposal number for each grant.

For research not funded by USDA, the Johns Hopkins University has developed a database to help locate currently or recently funded research projects funded by NSF, NIH, DOE and USDA. The text-searchable **Searching for Biologists database**[11] contains abstracts of the proposed work as well as investigator information, and the amount and date of award. For NIH and DOE supported work, the majority of entries relate to genome mapping. As an added feature, the database also includes a searchable e-mail address directory for crystallographers, protistologists, yeast researchers and registered participants in Biosci/Bionet electronic discussion groups.

One final database that covers current federal research is **FEDRIP** (Federal Research in Progress). The FEDRIP database is administered by the National Technical Information Service (NTIS), the 'one-stop shop' for federal science and technical information. The FEDRIP database provides advance information about some 150,000 federally-funded research projects. Agencies contributing to the FEDRIP database providing relevant forestry information include USDA, Department of Energy, Environmental Protection Agency, National Science Foundation as well as grants awarded by the Small Business Innovation Research (SBIR) programs supported by many research agencies. The database is accessible through on-line vendors (like Dialog®) and in CD-ROM format.

---

[10] The NRICGP Abstracts of Funded Research is available at:
gopher zeus.esusda.gov select USDA and Other Federal Agency Information/USDA: Agency Information/CSREES-NRI. NRI Competitive Grants Program OR at: http://WWW.ree.usda.gov/new/nri/nricgp.htm

[11] Grant Abstracts Searching: NSF, NIH, DOE and USDA - Searching for Biologists can be accessed via: *gopher* gopher.gdb.org   select Searching for Biologists.

An alternative source of federally-funded research projects is the Community of Science database which includes searchable full-text research summaries from NIH, NSF, SBIR and USDA.[11]

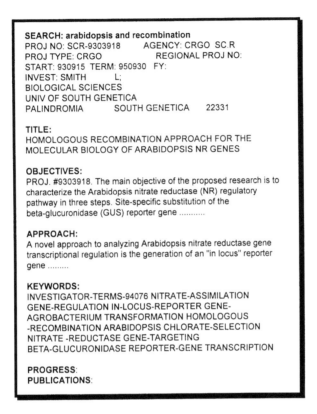

Figure 1: Example CRIS (Current Research Information System) record from Internet-based CRIS database.

Accessing the Agricultural Research Service (ARS)

The Agricultural Research Service (ARS) serves as the U.S. Department of Agriculture's primary in-house research agency. ARS spending in the area of biotechnology topped $113 million in 1993.[2] Several offices and tools are available to provide access to the research results in ARS labs. In addition to a National Technology Transfer Coordinator at ARS's main Beltsville, MD location, each of the ARS regional labs and individual research sites has a designated technology transfer contact. Technology transfer at ARS is a bottom - up process with individual researchers being responsible for getting the research used as well as done. To facilitate the transfer of technologies to the private sector ARS has developed several resources to improve communication between researchers and technology developers. Many of these tools can also be used by researchers to locate potential collaborators.

The **Technology Transfer Automated Retrieval System (TEKTRAN)** provides direct access to pre-publication research results and inventions available for licensing from the Agricultural Research Service (ARS) laboratories. Over 12,000 brief, non-technical interpretive summaries of the latest research results from ARS are available (see Figure 2). Most records also contain a technical abstract that provides more detail of the research. The database also contains information on more than 1,000 ARS inventions available for licensing. TEKTRAN is searchable by researcher, commodity or keyword and is updated monthly to include some 400 new summaries each month. Full access to the database is provided free of charge.[12] The reports contained in TEKTRAN can also be accessed via Internet with full-text search capabilities at one of the current sites.[13]

Title: INFLUENCE OF CRYSTAL PROTEIN COMPOSITION OF BACILLUS THURING ON CROSS RESISTANCE IN INDIAN MEAL MOTHS LEPIDOPTERA:PYRALIDA
subtitle: TEKTRAN
author: USDA Agricultural Research Service
date: 04/93

Interpretive Summary:
Bacillus thuringiensis is a biological insecticide that holds great promise as a replacement for chemical insecticides because of its safety for......

Technical Abstract:
The cross-resistance spectra of colonies of Plodia interpunctella (Hubner) resistant to strains of Bacillus thuringiensis (B.t.) subspp. kurstaki, aizawai, and entomocidus were compared. The cross-resistance spectra tended to reflect the toxin.........

*CONTACT
WM C MOTHMAN
1515 COLLEGE AVENUE   (approved 03/09/93)
FLATLAND, KS 61611
FLATLAND
KS 66502   913-776-1776   FTS none

Figure 2: Example of a TEKTRAN record from Internet-available TEKTRAN Reports

ARS also publishes a number of periodic reports that summarize the recent developments in ARS labs, including developments on cooperative research and development agreements (CRADAs), patent applications and licensing agreements. One such publication is the **Quarterly Report of Selected Research** which provides brief summaries of selected ARS research results for the preceding 3 months.

---

[12] For information about direct TEKTRAN access contact the National Technology Transfer Coordinator, ARS-USDA, BARC-West, 10300 Baltimore Blvd., Beltsville, MD 20705, USA, or by phone at 1-301-504-5345. Also available via Internet from the Technology Transfer Information Center (<http://WWW.nal.usda.gov/ttic>)

[13] ARS Research Reports are available via: *gopher* esusda.gov
select USDA and Other Federal Agency Information/ USDA: Agency Information/Agriculture Research Service (ARS) Reports, USDA OR searchable via *gopher* gopher.ces.ncsu.edu select National CES Information/ Search Research Results Database from USDA/.

The **Agricultural Research Information Letter** is a quarterly letter from the ARS Administrator to agribusiness executives that includes a condensed version of the quarterly report items which may be of particular interest to the commercial sector. A yearly review of research accomplishments is published in the **Research Progress in 199(5)**. All of these publications are available from the ARS Information Staff.[14]

Conclusions

While the federal government is the largest funding source for plant biotechnology, technology transfer efforts remain diffuse due to the fact that a number of funding agencies and funding mechanisms (intramural vs. extramural) are involved. Decentralized technology transfer efforts may in fact be very effective, yet the tools that are used to facilitate technology transfer, namely research information databases and networks, are most effective when they are standardized and easy to use. An inter-agency Technology Information Working Group recently surveyed biotechnology companies regarding the types of information found to be most useful for developing new technologies.[2] The results indicated that many of the resources, such as information about patents available for licensing, CRADA opportunities, and research summaries, are already available. But available and accessible are not necessarily synonymous terms.

New means of disseminating information rapidly and broadly (e.g., The Internet) has the potential to provide the accessibility and ease-of-use that may be lacking in the resources currently available. Given the increasing demands for research that is working toward "economic and societal" goals, future researchers will be best served by information systems that provide ready access to their research findings and improve their ability to be active participants in the technology transfer process.

---

[14] Contact the Information Staff, Agricultural Research Service - USDA, 6303 Ivy Lane, 4th Floor, Greenbelt, MD 20770, USA OR by phone at 1-301-344-2723/1-301-344-2824. Quarterly Reports are available at: *gopher* gopher.nal.usda.gov   select Other Ag. Publications/USDA, ARS, Quarterly Report.

From Lab Bench to Marketplace:
The Calgene FLAVR SAVR™ Tomato

Belinda Martineau

Calgene, Inc., 1920 Fifth Street, Davis, CA 95616, USA

Introduction

On May 21, 1994 a genetically engineered whole food, the FLAVR SAVR tomato, entered commerce for the first time. Since its original (re)generation, the FLAVR SAVR has been among the most well-studied tomatoes in history. It has received this scrutiny firstly because it has been at the forefront of plant science, representing one of the first examples of the successful use of antisense technology to inhibit endogenous plant gene expression. The FLAVR SAVR tomato has also been scientifically scrutinized because it has served as the "Plant Biotechnology Test Case" with the U.S. Department of Agriculture, the U.S. Food and Drug Administration, and consequently with American consumers. A historical perspective of the development of the FLAVR SAVR tomato will be presented in this review.

Scientific development

FLAVR SAVR tomatoes have highly reduced levels of the enzyme poly (1,4-α-D-galacturonide) glycan hydrolase (EC 3.2.1.15, polygalacturonase or PG) as compared to other commercially available tomatoes. Prior to the production of FLAVR SAVR tomatoes, several pieces of information in the scientific literature implicated the PG enzyme as a major contributor to tomato fruit softening during ripening. For example, several ripening mutants of tomato in which fruit-softening was delayed or decreased also contained deficient levels of PG enzyme activity (see, for example, Tigchelaar et al, 1978). Also, PG enzyme activity (Hobson, 1965) and protein abundance (Tucker et al, 1980) [and later mRNA accumulation (DellaPenna et al, 1987)] were all found to increase with a time course similar to that observed for fruit softening during tomato ripening. *In vitro* studies also demonstrated that PG could dissolve the middle lamella of green tomato fruit cell walls (DellaPenna et al, 1987) and solublize pectin in cell wall preparations isolated from green tomato fruit (Wallner and Bloom, 1977; Huber, 1983) as would be expected of an enzyme responsible for carrying out these softening-related phenomena in ripe fruit. Based in large part on these correlative data, scientists at Calgene set out to clone the PG gene from tomato using recombinant DNA technology.

Although the PG enzyme is abundant in ripe tomato fruit, the strategy chosen for reducing PG levels utilized antisense technology. The method was far from proven in plant systems at that time. It had been reported that an excess of antisense RNA was often required for effective reduction of target mRNA levels, and reliable methods for the efficient introduction of genes into tomato and subsequent regeneration of transformed tomato cells into plants were also not available in 1984.

Sheehy and co-workers (Sheehy et al, 1987) at Calgene reported the successful isolation of a PG cDNA clone, using an immunological detection method. The isolated PG cDNA clone was used to produce chimeric gene constructs designed to produce antisense PG RNA when expressed in plant cells. Using the tomato transformation/regeneration methods worked out by Fillatti and co-workers at Calgene (Fillatti et al, 1987), tomato plants expressing antisense PG RNA were produced and characterized. PG mRNA, protein and enzyme activity levels were successfully reduced in ripe fruit of the transformed plants (Sheehy et al, 1988). A reduction of more than 95% in PG activity was recorded for some primary transformants, with greater than 99% reduction observed in fruit homozygous for the antisense PG gene. Production of the normally highly abundant PG enzyme had been successfully inhibited through the use of antisense technology.

Field trials of tomato plants exhibiting highly reduced levels of PG were then conducted. Because seed homozygous for the chimeric antisense PG gene was only available in time for the winter season 1988/1989, the first Calgene field trial of genetically engineered antisense PG tomato plants took place in Guasave, Sinaloa, Mexico. Fruit from this trial was analyzed primarily for its field-holding ability and the quality of its processed juice. Kramer and co-workers (Kramer et al, 1990) reported that in a genetically engineered tomato line which expressed 8% of normal PG activity, the percentage of rotten fruit was significantly less in the last three of five harvests (the first of which was conducted when about 95% of the fruits were ripe) than in the non-transformed control variety. Processed juice from the same line had significantly greater viscosity than did juice from the control parental line.

In subsequent field trials conducted in California and Florida, USA, fresh market and processing varieties of FLAVR SAVR tomatoes were analyzed for their shelf life, firmness, and resistance to various tomato fungal pathogens, among other traits (Kramer et ɛl. 1992). Shelf life was dramatically increased in FLAVR SAVR tomatoes as compared to fruit from non-transgenic control plants. This increase in shelf life can be most dramatically demonstrated by way of illustration, such as is shown in Figure 1. The extended life of FLAVR SAVR tomatoes was also utilized to increase fruit time on the vine and consequently improve tomato flavor.

Figure 1: Quality of transgenic and control fresh market tomatoes after extended storage. Mature green non-transformed (left) and antisense PG (right) fruit were harvested and packed into 25-lb. boxes, treated with ethylene, and stored for a total of 25 days. After storage, fruit were removed from randomly selected boxes for comparison as shown. (Reprinted from Kramer et al, 1992 with permission.)

FLAVR SAVR tomatoes also exhibited improved firmness as compared to fruit from appropriate non-transgenic controls harvested at the same stage of ripeness (Kramer et al, 1992). The improvement in fruit firmness that resulted from the expression of antisense PG RNA was more dramatic in some genetic backgrounds than in others. This fact, coupled with differences in (and subtleties of) firmness measuring techniques, contributed to initial reports by another group (Smith et al, 1988; Schuch et al, 1991) that no change in the softening of antisense PG transgenic fruit was observed. To date, however, results from independent scientific teams have confirmed the Calgene firmness finding in fruit expressing antisense PG RNA (see, for example, Grierson and Schuch, 1993).

A significant improvement in resistance against fungal pathogens of ripe tomato fruit has also been documented in FLAVR SAVR tomatoes. Kramer et al (1992) measured the area of infection caused by the pathogens responsible for tomato soft rot (*Rhizopus stolonifer*) and sour rot (*Geotrichum candidum*) in fruit expressing antisense PG RNA and in control non-transformed fruit. They found statistically, and dramatically, less pathogen ingress in the transgenic fruit. To further corroborate that the improvement in pathogen resistance was due to reduced levels of PG in the transgenic fruit, an additional study was conducted (Kramer et al, 1992). The area of infection caused by *G. candidum* on fruit from a tomato mutant called Neverripe (*Nr*), and on fruit from an *Nr* plant into which a copy of the PG gene in the sense orientation had been transferred, was also measured. As the name implies, *Nr* fruit normally fail to ripen. *Nr* fruit also fail to soften and are deficient in PG. The results of the study by Kramer and co-workers indicated that *Nr* fruit are also quite resistant to *G. candidum*. However, when, as the result of expression of the introduced sense PG gene *Nr* fruit produced significant amounts of PG enzyme, susceptibility to *G. candidum* was dramatically increased (Kramer et al, 1992). As PG affects the integrity of tomato fruit cells, it also apparently affects the susceptibility of the fruit to fungal pathogens.

Regulatory development

USDA

Permits had been obtained from the U.S. Department of Agriculture (USDA) for each of eight contained field trials of FLAVR SAVR tomatoes that had been conducted since 1989. In general, no changes that might affect the weediness potential of FLAVR SAVR tomato plants were observed during these trials; FLAVR SAVR tomatoes had similar horticultural traits as traditionally bred tomatoes. For each trial, the USDA conducted a thorough review, prepared an environmental assessment, and found that field testing of FLAVR SAVR tomatoes had no significant impact on the environment. In each case the USDA concluded that "a determination was made that this limited field trial does not pose a risk of introduction or dissemination of a plant pest and does not present a significant impact on the quality of the human environment."

With this field trial track record, and a substantial package of supporting data, Calgene filed a petition in May, 1992 requesting that USDA determine that FLAVR SAVR tomatoes do not present a plant pest risk, are not otherwise deleterious to the environment, and are therefore not a regulated article under USDA Animal and Plant Health Inspection Service (APHIS) regulations. Based on the data submitted and extensive review by the agency, the USDA issued a determination in October, 1992 declaring that FLAVR SAVR tomatoes, which had previously been field tested under USDA regulations, "will no longer be considered regulated articles under APHIS regulations at 7 C.F.R. §340. Permits under those regulations will no longer be required from APHIS for field testing, importation, or interstate movement of those tomatoes or their progeny." This determination was the first such approval for growing a genetically engineered crop uncontained and using the same procedures as any other traditionally bred crop.

FDA: selectable marker gene and gene product

Efforts to gain official approval from the U.S. Food and Drug Administration (FDA) for commercial sale of FLAVR SAVR tomatoes started in 1989 (for an additional review see Redenbaugh et al, 1994). No requirements to attain such approval for genetically engineered plant products were in place at that time. The Calgene effort was therefore proactive and embarked upon for several reasons, not the least of which was related to educating the public about agricultural biotechnology.

The first Calgene request, submitted November 26, 1990, was for an advisory opinion from FDA on the use of the selectable marker gene, *kan^r*, and its gene product, neomycin phosphotransferase (APH(3')II), in transgenic tomato, cotton and rapeseed plants (Calgene, 1990). Approval was requested for the use of *kan^r* in all three crops in anticipation of the introduction by Calgene of additional genetically engineered plant varieties and plant products including cotton with the BXN™ (bromoxynil resistant) gene conferring tolerance of the crop to the herbicide bromoxynil for improved weed control; cotton resistant to specific insect pests such as the cotton bollworm and tobacco budworm; and rapeseed varieties that will produce specialty edible and industrial oils.

Some of the major issues addressed in this document included 1) horizontal gene transfer, could the *kan^r* gene move from transgenic plants to other organisms and if so, what would be the impact of such movement; 2) is the gene product, APH(3')II, a toxin or allergen; and 3) could APH(3')II compromise an oral dose of kanamycin or neomycin in human antibiotic therapy?

This request for an advisory opinion was submitted in a Food Additive Petition (FAP) format because there was a potential that Calgene might choose an FAP for APH(3')II after subsequent consultations with the FDA. The submission was, in fact, changed to a request of FDA to issue a food additive regulation for APH(3')II, encoded by the *kan^r* gene and used as a processing aid in the production of transgenic plants, on January 4, 1993. One reason this change was made was to send a clear message to American consumers that FDA had conducted a standard and thorough safety review.

Calgene's approach to demonstrating the safety of the APH(3')II selectable marker was partly based on the recommendations of the International Food Biotechnology Council (IFBC, 1990). The IFBC considered that the probability of horizontal gene flow from transgenic plants was vanishingly small. In support of that conclusion the Calgene document contained detailed calculations, made using worst-case scenarios, to estimate the potential increase in the number of human gut or soil bacteria that could become resistant to kanamycin should horizontal gene transfer occur as a result of consuming or growing, respectively, FLAVR SAVR tomatoes. The results of these calculations indicated that there would be no significant increase in kanamycin- or neomycin-resistant soil or human gut bacteria resulting from the commercial growing or consumption, respectively, of tomatoes containing the *kan^r* gene. For example, under worst case and extremely rare conditions, the potential increase in the number of human gut bacteria that could become resistant to kanamycin is 0.00000000000026%. This means that for every 380 humans that consume the mean level of genetically engineered tomatoes, one gut bacterium susceptible to kanamycin may become resistant to the antibiotic (see also Redenbaugh et al, 1993). This number is insignificant when compared to the number of kanamycin-resistant bacteria that occur naturally in the human gut and in the environment (see, for example, Anderson et al, 1987 and Van Elsas and Pereira, 1986).

Some of the naturally occurring kanamycin-resistant human gut bacteria produce APH(3')II. This fact provided some evidence that this protein was not toxic to humans. APH(3')II is also not homologous to any known toxins or allergens, and in fact, has been used in human gene therapy (Kasid et al, 1990). APH(3')II is also not known to have any properties that distinguish it toxicologically from any other phosphorylating enzyme, many of which are known to be present in the food supply. Glycosylation and subsequent increase in the antigenic capacity of APH(3')II could not occur because APH(3')II does not contain the necessary sequence information for transport to the subcellular locations at which glycosylation reactions take place. These facts, also noted in recent peer-reviewed articles (Flavell et al, 1992; Nap et al, 1992), in addition to experimental evidence that APH(3')II is degraded and inactivated in the human gastrointestinal tract to a level that does not represent a risk of toxicity, served as the basis for Calgene's conclusion that this protein is neither a toxin nor an allergen.

The experiments conducted at Calgene showing that APH(3')II, as is the case for any other typical protein, was inactivated by pepsin in simulated gastric fluids and by simulated intestinal fluids, also provided support for arguments regarding the issue of possible compromised efficacy of oral therapeutic use of kanamycin and neomycin due to the presence of the enzyme in FLAVR SAVR tomatoes. Even if not degraded, it was argued in an additional document submitted by Calgene October 30, 1992, APH(3')II would be inactive in the absence of the energy producing cofactor ATP and under the low pH conditions of the human gut. The results of this evaluation demonstrated that only a small fraction of neomycin could be inactivated under a worst-case scenario.

On May 18, 1994, the FDA concluded that "the proposed use of APH(3')II as a processing aid in the development of new varieties of tomato, oilseed rape, and cotton is safe". FDA also "determined that there is no need to set a tolerance for the amount of APH(3')II that will be consumed. . . " and concluded that "FDA agrees with Calgene that the characteristics of APH(3')II do not raise a safety concern (Federal Register 59:26700)."

FDA: FLAVR SAVR™ tomato

Calgene submitted an extensive data package to the FDA in August 1991 covering issues related specifically to the FLAVR SAVR tomato and requesting that the agency consider the information and provide an Advisory Opinion with regard to the product's safety (Docket #91A-0330/API). The information Calgene provided demonstrated that FLAVR SAVR tomato varieties were nutritionally and compositionally indistinguishable from non-transformed control lines of the same genetic background except in terms of characteristics related to pectin and the presence of the novel *kan*^{r} gene and APH(3') products. Levels of nutrients, taste, horticultural and developmental traits (yield, disease resistance, plant morphology, etc.) and levels of potential toxins (solanine and tomatine) were unchanged. The data also indicated that no adverse pleiotropic traits were detected in FLAVR SAVR tomatoes selected for low PG activity and horticultural characteristics.

The August 1991 submission also contained an extensive molecular analysis of the DNA that had been inserted into FLAVR SAVR tomato plants. Using the endogenous PG gene as an internal standard (Sanders et al, 1992), DNA blot hybridizations were used to determine the number of antisense PG genes inserted into particular FLAVR SAVR lines and to demonstrate that those inserted genes behaved in a Mendelian manner in subsequent generations. These data were important in establishing the precision of the techniques of agricultural biotechnology. The demonstration of Mendelian inheritance also supported the claim that once the plants were produced, the DNA in FLAVR SAVR tomatoes "behaved" like DNA in any other tomato; it was just as stable and segregated among progeny plants similarly. The DNA blot hybridization evidence also determined that the antisense PG genes had been inserted into only one genetic locus in the FLAVR SAVR plants examined. These results minimized the possibility of pleiotropic effects caused by disruption of endogenous genes during the transformation process.

In May of 1992 FDA released its Statement of Policy regarding genetically engineered plant products and concluded that the potential for pleiotropy, i.e. unexpected effects, was extremely low in food plants with a long history of breeding. The Statement of Policy noted, however, that animal feeding studies could be a means of addressing such potential. The FDA suggested that conducting animal feeding studies with the FLAVR SAVR tomato could confirm the absence of pleiotropic effects and validate the compositional analysis approach to safety assessment. The decision was made at Calgene that the most thorough safety analysis possible should be conducted on the FLAVR SAVR due to its unique status as the first genetically engineered whole food to be available to the public. The company therefore had three 28-day oral (intubation) safety studies conducted with rats.

No significant toxicological effects or other unexpected major treatment effects were noted in the intubation studies (Redenbaugh et al, 1994, for more details). Included in this submission was the opinion of an expert panel of food safety scientists, who concluded that the FLAVR SAVR tomato was as safe as any other tomato. The results of the animal intubation studies also validated the molecular, chemical and compositional approach that had been used to demonstrate the safety of the FLAVR SAVR tomato, and confirmed that there was no basis for a safety concern.

Additional data supplied to the FDA in response to letters from the Agency of 2 December 1992 and 8 January 1993 included a complete characterization, with regard to potential open reading frames and any potential corresponding proteins, of all the DNA sequences within the T-DNA region in FLAVR SAVR tomatoes. FDA also requested, and was provided with, data demonstrating that FLAVR SAVR tomatoes, stored to the end of their shelf life, had levels of provitamin A and vitamin C that were within the normal range for these vitamins in non-transgenic varieties of tomatoes. The FDA also asked in a letter of 6 April 1993 for experimental evidence demonstrating that only the DNA in the T-DNA region of the Ti plasmid was integrated into the tomato genome. Pre-commercial FLAVR SAVR lines examined using the Southern blot technique showed no hybridization to vector DNA sequences other than to those within the T-DNA region. Interestingly, evidence for vector sequences from outside the T-DNA region was evident in other, non-FLAVR SAVR, transgenic plants examined (Martineau et al, 1994).

Calgene's final data submissions to the FDA were in response to the Agency's letter of 8 June 1993 regarding the level of glycoalkaloids in tomatoes and methods used for their measurement. Glycoalkaloids, particularly solanine, had long been a safety concern in potatoes since improper tuber storage or injury can result in dangerously high levels of these compounds. Most glycoalkaloid studies had consequently been conducted in potato. In tomato, the most abundant glycoalkaloid, tomatine, was known to be rapidly degraded during the fruit ripening process. Since fruit safety had therefore not been an issue very few measurements of tomatine had been previously conducted in tomato. The lack of concern regarding glycoalkaloids in tomato was evident in the paucity of related reports in the scientific literature and it translated into a lack of current, state of the art methods for measuring these compounds in tomato fruit. To rectify this situation, a high-performance liquid chromatography protocol for α-tomatine analysis was optimized at the University of Maine by Rodney Bushway (unpublished) and utilized to analyze FLAVR SAVR tomatoes. Only 5 out of 98 red tomato samples (four control and one transgenic fruit) had detectable, low amounts of tomatine. Detectable quantities of additional glycoalkaloids, solanine and chaconine, were not found. This information on glycoalkaloid levels and measurements comprised the last data sent to FDA as part of Calgene's safety assessment of the FLAVR SAVR tomato. It was submitted on 1 October 1993.

FDA held a Food Advisory Committee meeting on April 6-8, 1994, at which the FDA Center for Food Safety and Applied Nutrition (CFSAN) concluded that the FLAVR SAVR tomato was as safe and nutritious as other commonly consumed tomatoes and "that the use of the *kan^r* gene and APH(3')II will not have a significant impact on the environment and that an environmental impact statement is not required." On May 18, 1994 the FDA finished its safety review of the FLAVR SAVR tomato and APH(3')II. The FDA concluded that "FLAVR SAVR™ tomatoes have not been significantly altered when compared to varieties of tomatoes with a history of safe use" (Federal Register 59:26646) and "that the use of aminoglycoside 3'-phosphotransferase II is safe for use as a processing aid in the development of new varieties of tomato, oilseed rape and cotton intended for food use" (Federal Register 59:26700).

Business development

Three days after FDA had issued its official approval of the FLAVR SAVR, the first genetically engineered whole food was available for sale to consumers. Calgene was able to get FLAVR SAVR tomatoes to market so quickly because Calgene Fresh, Inc. had been established in January 1992 to meet production and marketing needs for the new product. Calgene Fresh had been preparing for sales of genetically engineered fruit by selling non-transgenic, vine-ripened tomatoes since April 1992. During the intervening two years various aspects of the FLAVR SAVR business strategy had been tested and the brand name MacGregor's® had been introduced. Although not required to do so by law, point of purchase information describing the production of FLAVR SAVR tomatoes, including the use of the selectable marker gene as a processing aid, had also been prepared. Relationships with growers, packing facilities, production crews and sales staff were all in place by the time FDA approval was obtained.

The company had also secured its position in the fresh market tomato business by way of licensing agreements and in terms of patent protection. Calgene holds issued patents covering the PG gene sequence and its use in plants (Hiatt et al, 1989) and the use of antisense technology to regulate gene expression in plants (Shewmaker et al, 1992). The test market carried out by Calgene Fresh with non-transgenic fruit also served to validate concepts developed from consumer marketing information. That information had indicated that consumers were unhappy with the quality of fresh market tomatoes picked green and treated with ethylene gas and were willing to pay a premium price for premium quality fruit. Sales of non-transformed MacGregor's brand tomatoes from 1992 to 1994 established the true value of a high quality, vine-ripened tomato in the eyes of the consumer. Since the sale of the first FLAVR SAVR tomato demand for the genetically engineered fruit has been high.

Conclusions

This paper presented a brief history and overview of the development of the first whole food developed using genetic engineering, the FLAVR SAVR tomato. Transfer of the technologies involved in producing the FLAVR SAVR tomato from the laboratory to the marketplace has been accomplished. The stage is now set for other genetically engineered plant varieties and plant products to follow suit.

References

Anderson, W.F., Blaese, R.M., Nienhuis, A.W. & O'Reilly, R.J. (1987) Submitted to the Human Gene Therapy Subcommittee. Recombinant DNA Advisory Committee. National Institutes of Health. Bethesda, Maryland. April 24.

Calgene (1990) Request for Advisory Opinion, U.S. Food and Drug Administration, Docket #90A-0416.

DellaPenna, D., Kates, D. & Bennett, A. (1987) *Plant Physiol.* **85**, 502-507.

Fillatti, J.J., Kiser, J., Rose, R. & Comai, L. (1987) *Bio/Techenology* **5**, 726-730.

Flavell, R., Dart, E., Fuchs, R. & Fraley, R. (1992) *Bio/Technology* **10**, 141-142.

Grierson, D. & Schuch, W. (1993) *Philosophical Transactions of the Royal Society of London* **342**, 241-250.

Hiatt, W.R., Sheehy, R.E., Shewmaker, C.K., Kridl, J.C. & Knauf, V. (1989) U.S. Patent Number: 4,801,540.

Hobson, G. (1965) *J. Hort. Sci.* **40**, 66-72.

Huber, D.J. (1983) *J. Am. Soc. Hort. Sci.* **108**, 405-409.

IFBC (International Food Biotechnology Council) (1990) *Regul. Toxicol. Pharmacol.* **12**, S1-S196.

Kasid, A., Morecki, S., Aebersold, P., Cornetta, K., Culver, K., Freeman, S., Director, E., Lotze, M.T., Anderson, W.F. & Rosenberg, S.A. (1990) *Proc. Natl. Acad. Sci. USA* **87**, 473-477.

Kramer, M., Sanders, R.A., Sheehy, R.E., Melis, M., Kuehn, M. & Hiatt, W.R. (1990) In: *Horticultural Biotechnology*, Wiley-Liss, Inc. pp 347-355.

Kramer, M., Sanders, R., Bolkan, H., Water, C., Sheehy, R.E. & Hiatt, W.R. (1992) *Posthar. Biol. Tech.* **1**, 241-255.

Martineau, B., Voelker, T. & Sanders, R.A. (1994) *Plant Cell* **6**, 1032-1033.

Nap, J.-P., Bijvoet, J. & Stiekema, W. (1992) *Transgenic Res.* **1**, 239-249.

Redenbaugh, K., Hiatt, W.R., Martineau, B., Kramer, M., Sheehy, R., Sanders, R., Houck, C. & Emlay, D. (1992) Safety Assessment of Genetically-Engineered Fruits and Vegetables: A Case Study of the FLAVR SAVR^{TM} Tomatoes, CRC Press, Boca Raton, p. 267.

Redenbaugh, K., Berner, T., Emlay, D., Frankos, B., Hiatt, W., Houck, C., Kramer, M., Malyj, L., Martineau, B., Rachman, N., Rudenko, L., Sanders, R., Sheehy, R. & Wixtrom, R. (1993) *In Vitro Cell. Dev. Biol.* **29P**, 17-26.

Redenbaugh, K., Hiatt, W., Martineau, B. & Emlay, D. (1994) *Trends Food Sci. Tech.* **5**, 105-110.

Sanders, R.A., Sheehy, R.E. & Martineau, B. (1992) *Plant Molec. Biol. Rep.* **10**, 164-172.

Schuch, W., Hobson, G., Kanczler, J., Tucker, G., Robertson, N., Grierson, D., Bright, S. & Bird, C. (1991) *HortSci.* **26**, 1517-1520.

Sheehy, R.E., Pearson, J., Brady, C.J. & Hiatt, W.R. (1987) *Mol. Gen. Genet.* **208**, 30-36.

Sheehy, R.E., Kramer, M. & Hiatt, W.R. (1988) *Proc. Natl. Acad. Sci. USA* **85**, 8805-8809.

Shewmaker, C.K., Kridl, J.C., Hiatt, W.R. & Knauf, V. (1992) U.S. Patent Number: 5,107,065.

Smith, C.J.S., Watson, C.F., Ray, J., Bird, C.R., Morris, P.C., Schuch, W. & Grierson, D. (1988) *Nature* **334**, 724-726.

Tigchelaar, E., McGlasson, W. & Buescher, R. (1978) *HortSci.* **13**, 508-513.

Tucker, G., Robertson, N. & Grierson, D. (1980) *Eur. J. Biochem.* **112**, 119-124.

Van Elsas, J.D. & Pereira, M.T.P.R.R. (1986) *Plant and Soil* **94**, 213-226.

Wallner, S.J. & Bloom, H.L. (1977) *Plant Physiol.* **60**, 207-210.

Biotechnological Applications of Inheritable and Inducible Resistance to Diseases in Plants

Sadik Tuzun, Peter A. Gay, Christopher B. Lawrence,
Tracy L. Robertson and Ronald J. Sayler

*Department of Plant Pathology and Biological Control Institute, Auburn University,
Auburn, AL 36849-5409,USA*

Introduction

Plants have developed a broad array of responses to pathogens, which are categorized into three distinct groups. A non-host interaction is characterized by the pathogen being incapable of replication, and the plant does not suffer from the interaction. Most interactions between plants and phytopathogens are of this type. An incompatible interaction occurs when the pathogen is capable of establishing an initial infection, but the plant is able to defend itself via a wide range of cellular processes resulting in an integrated set of resistance responses. A compatible interaction occurs when a pathogen establishes an infection in the plant and may translocate systemically. Manifestation of disease generally occurs through visual symptoms which may range from mild to severe. In a compatible interaction, the host response to pathogenic attack may be delayed or may not occur at all. Under certain conditions, however, a compatible interaction can be converted to an incompatible one through the activation of latent defensive responses. This phenomenon is called "induced systemic resistance" (ISR) or "systemic acquired resistance" (SAR). Understanding the mechanisms of disease resistance achieved either through breeding or induced by abiotic or biotic agents may lead to safer and more effective approaches to disease control. Such mechanisms may be manipulated to increase their effectiveness by utilization of biotechnological approaches. In this chapter, we will briefly discuss the nature of disease resistance in plants and then focus specifically on the role of pathogenesis-related (PR) proteins in disease resistance. This discussion will mainly concentrate on the current research in our laboratory in three plant-pathogen systems. The overall focus of our research is to determine the role of specific chitinase isozymes associated with disease resistance.

Concepts of disease resistance

The susceptible reaction to disease usually occurs when the pathogen becomes established in the plant and multiplies sexually or asexually. Reproduction allows further infection of the same host or other susceptible hosts. Resistance to disease is broadly defined as the active state of the host which creates unsuitable conditions for ingress, establishment, and reproduction of the pathogenic organisms. Resistance to disease in plants has been recognized over centuries through the identification of wild species which survived in the presence of pathogens. Qualitatively-inherited resistance is controlled by one or a few genes, while quantitatively-inherited resistance may be due to the presence of multiple genes. Either of these types of resistance may result in the activation of physiological and molecular defense mechanisms in the host. However, these mechanisms may be overcome by the pathogen through adaptation and development of more virulent races. Breeders have been somewhat successful in developing cultivars with a broad spectrum of disease resistance, but this is often difficult while maintaining other agronomically important traits.

Research in many laboratories indicates that, although the plant is susceptible, it still contains the genetic potential for resistance mechanisms to multiple (bacterial, fungal and viral) pathogens (Kuc´, 1985; Tuzun and Kuc´, 1991; Tuzun and Kloepper, 1994). Activation of these latent resistance mechanisms will lead to resistance in a previously susceptible plant. This phenomenon of induced resistance is defined by Kloepper et al (1992) as 'the process of active resistance dependent on the host plant's physical or chemical barriers, activated by biotic or abiotic agents'. Induced resistance has been documented in at least 26 agronomically important crops (Tuzun and Kuc´, 1991) including cucumber (Wei et al, 1991), tomato (Heller and Gessler, 1986) and tobacco (Tuzun and Kuc´, 1985). It is important to note that induced resistance protects plants against multiple organisms and this indicates that the resistance mechanisms may involve multiple genes.

Plant defense responses can be induced by pathogenic and non-pathogenic organisms as well as various chemicals (Tuzun and Kloepper, 1994). Non-detrimental biological inducers of resistance have gained increasing attention due to the fact that they satisfy environmental concerns while chemical or abiotic inducers may not. One bio-intensive method of protecting plants involves 'bacterializing' seeds or roots of plants with select strains of plant growth-promoting rhizobacteria (PGPR) which protect plants against soil-borne and foliar pathogens either by competition or by systemically inducing resistance. Wei et al (1991) found that six of ninety-four strains of PGPR systemically induced resistance in cucumber to the anthracnose pathogen *Colletotrichum orbiculare* upon seed treatment. Root colonization with *Pseudomonas fluorescens* strain WCS417r has provided protection against the stem-inoculated pathogen *Fusarium oxysporum* f. sp. *dianthi* in carnation (van Peer et al, 1991). Inoculation of common bean seeds with *P. fluorescens* (S97) induced systemic resistance to *P. syringae* pv. *phaseolicola*, the halo blight pathogen (Alstrom, 1991). Strain CHA0 of *P. fluorescens* systemically induced resistance in tobacco to Tobacco Necrosis Virus (TNV) upon soil inoculation (Maurhofer et al, 1994). In all the above mentioned studies, the protection of plants appears to be related to the changes in host defense responses. Indeed, an increase in peroxidase activity in immunized cucumbers (Wei et al, 1992), accumulation of phytoalexins in immunized carnations (van Peer et al, 1991), accumulation of pathogenesis related (PR) proteins in common bean (Hynes and Lazarovits, 1989) and an increase in PR-1 group proteins, β-1,3-glucanases and chitinases in the intercellular fluids of protected tobacco leaves (Maurhofer et al, 1994) was observed.

Mechanism(s) of disease resistance

Current understanding of resistance at the genetic level includes the existence of 'responsive' genes in the host and pathogen that are expressed during plant-pathogen interactions. This equates to a physiological response in the plant upon recognition of the pathogen. Following this recognition event, plants generally respond through the induction of several local responses in the cells immediately surrounding the infection site. These include a localized cell death, known as the hypersensitive response which is characterized by the deposition of callose (Kauss, 1987), physical thickening of cell walls by lignification (Vance et al, 1980), and the synthesis of vacuolar antibiotic molecules (Dixon, 1986) and proteins, such as cell wall hydrolases (Bowles, 1990).

Gene-for-gene interactions, in which disease resistance involves a single resistance (R) gene in the plant corresponding to a single avirulence (*avr*) gene in the pathogen, have been described in many plant-pathogen model systems (Kobayashi et al, 1989; Meshi et al, 1989; Keen and Buzzell, 1991; Debener et al, 1991; Dong et al, 1991). Genetic factors in both the host and the pathogen are involved in determining the specificity of these responses and thus, determining the nature of the resistance. Because of the lack of knowledge about the products of R genes, their isolation and characterization has been difficult.

Recently, the *Pto* gene from tomato has been cloned and characterized (Martin et al, 1993). This gene confers resistance to races of *P. syringae* pv. *tomato* which carry the avirulence gene *avrPto*. When susceptible tomato plants were transformed with the *Pto* cDNA, they showed resistance to the pathogen. Analysis of the amino acid sequence revealed a similarity to serine-threonine protein kinases, suggesting a role for *Pto* in a signal transduction pathway. The chromosomal location of resistance genes to other pathogens has been mapped in their respective host species. Transposon tagging and/or map-based cloning have been used to locate the dominant *N* locus of tobacco which confers resistance to TMV (Whitham et al, 1994), the *Cf*-genes of tomato which confer resistance to specific races of *Cladosporium fulvum*, (Jones et al, 1994) and *RPS2*, a resistance locus located in *Arabidopsis* to *P. syringae* (Bent et al, 1994; Mindrinos et al, 1994). The *rps2* gene is characterized by leucine-rich repeats as well as a nucleotide binding site, all characteristic of proteins involved in the signal transduction pathway.

Research has indicated that a specific group of proteins called pathogenesis related (PR) proteins accumulate in infected plants (van Loon and van Kammen, 1970 and Gianinazzi et al, 1970), and their accumulation patterns correlated with disease resistance (Tuzun and Kuc', 1991). PR proteins are characterized by being relatively low molecular weight, selectively extractable at low pH, and resistant to proteolytic digestion (van Loon, 1985). The induction of PR proteins by pathogen attack or elicitor treatment has been documented in several cases (Gianinazzi et al, 1970; van Loon and van Kammen, 1970; Kauffman et al, 1987; Tuzun et al, 1989). Research repeatedly has shown the induction of systemic resistance results in an increase of both chitinases and β-1,3-glucanases in many different plant systems such as cucumber (Kuc', 1987), tobacco (Tuzun et al, 1989) and tomato (Cohen et al, 1994). Therefore, they are thought to play an important role in active host resistance. PR proteins are designated from PR-1 to PR-5, based on their relative mobility in native gels, with each class consisting of both acidic and basic isoforms (for review see Linthorst, 1991). PR-1 proteins are low molecular weight, approximately 15,000 Daltons, and consist of both basic (PR-1b) and acidic (PR-1a) isoforms. Their function is yet unknown, however, transgenic tobacco overexpressing the PR-1 gene are more resistant to infection by *Peronospora tabacina* than control tobacco plants. PR-2 and PR-3 group proteins have been identified as β-1,3-glucanases and chitinases, respectively. Many plant pathogenic fungi contain β-1,3-glucans and chitin in their cell walls making them subject to hydrolytic attack, especially at the hyphal tip (Schlumbaum et al, 1986). Also, many plant chitinases have lysozyme activity allowing them to degrade the peptidoglycan layer of bacteria (Boller, 1985). PR-4 proteins are of unknown function, while PR-5 proteins are thaumatin-like, amphipathic in character and are believed to disrupt membrane permeability (Bol et al, 1990; Linthorst et al, 1991).

Specific chemicals such as salicylic acid (SA) have been hypothesized to be endogenous regulators of local and systemic disease resistance and inducers of pathogenesis-related (PR) proteins (Yalpani et al, 1993). The application of SA increases the resistance in a number of plant species to viral, bacterial and fungal pathogens (Mills and Wood, 1984; White, 1979; Ye et al, 1989). Increases in the levels of endogenous SA were observed to correlate with the expression of defense-related genes and the development of ISR (Malamy et al, 1992).

Although the role of SA is still somewhat unclear, transgenic tobacco plants which overexpress salicylate hydroxylase were defective in their ability to induce systemic resistance to TMV (Gaffney et al, 1993).

Induction of chitinase/lysozyme isozymes in relation to disease resistance

Chitinases catalyze the hydrolysis of chitin, a linear polymer of β-1,4-linked *N*-acetylglucosamine residues, which is the predominant constituent of fungal cell walls, nematode eggs, and mid gut layers of insects. Chitinases have been found in bacteria, fungi, nematodes, and plants (Hennis and Chet, 1975), and are known to play a role in growth and extension of fungal hyphae. Since chitin is not found in plants, it has been proposed that plant chitinases play a role in defense against invading pathogens. In addition, some plant chitinases exhibit lysozyme activity which may be indicative of antibacterial activity (Boller, 1985).

Chitinases represent a highly variable group of enzymes that share similar catalytic activity. Three classes of plant chitinases have been proposed based on protein primary structure (Shinshi et al, 1990). Class I chitinases have an N-terminal cysteine-rich domain of approximately 40 amino acids and a highly conserved main structure, with the two regions separated by a variable hinge. Class II chitinases lack the N-terminal cysteine-rich domain but have high amino acid sequence similarity to the main structure of class I chitinases (Linthorst et al, 1990). Class III chitinases have no sequence similarity to the class I or II enzymes (Lawton et al, 1992). The highly variable nature of chitinases suggests that isozymes of chitinase may carry out specific roles.

Plants have multiple chitinase isozymes and recent evidence suggests that only a few chitinase isozymes may have antifungal activity. This antifungal activity is specific for certain pathogens (Sela-Buurlage et al, 1993). Our research group has been investigating whether plants bred for resistance to diseases express specific enzymes earlier and to a greater extent than susceptible lines. We have also been investigating differences in the accumulation patterns of chitinase isozymes in relation to both inducible and inheritable disease resistance in three plant-pathogen systems: tobacco-*P. tabacina*, tomato-*Alternaria solani*, and cabbage-*Xanthomonas campestris* pv. *campestris* (Xcc). The following sections will focus on the research we have been conducting to elucidate the possible role of chitinases in disease resistance, and related studies on these three plant-pathogen systems.

Tobacco

Tobacco has been used as a model system to elucidate mechanisms of induced systemic resistance for many years (Tuzun and Kuc´, 1989). Resistance produced in this manner is effective against a broad range of pathogens, in addition to the pathogen which triggered the initial response. For example, inoculation of tobacco with TMV leads to induced systemic resistance not only to TMV but to *P. tabacina* and *Phytophthora parasitica* var. *nicotianae* as well (McIntyre et al, 1981). Induced systemic resistance to *P. tabacina*, the causal agent of blue mold disease, results in the accumulation of phytoalexins (Stolle et al, 1988), salicylic acid (Yalpani et al, 1993) and many PR proteins, including chitinases and β-1,3-glucanases (Tuzun et al, 1989). The role of chitinases in resistance to *P. tabacina* may be a more generalized defense response since the pathogen's fungal cell wall is primarily composed of β-1,3 linked glucans. However, the chitinase isozymes appear to be regulated differentially during the establishment of induced systemic resistance. Pan and his colleagues (1992) detected an increase in the activities of two chitinase isozymes of eight total in tobacco plants systemically protected against *P. tabacina*. These two chitinase isozymes were located predominantly in the intercellular spaces which is the primary area of infection for this pathogen.

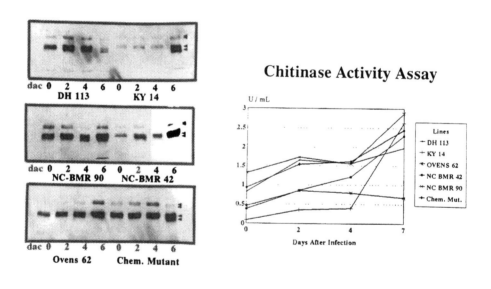

Figure 1: Western blot analysis (left) and enzyme activity assay (right) of tobacco chitinases. Acid-extractable proteins from tobacco lines resistant and susceptible to P. tabacina *were blotted onto Nylon membrane and probed with antibodies raised against tobacco chitinases. The chitinase enzyme activity assay was performed utilizing RBV-chitin as substrate. Breeding lines DH 113, Ovens 62, NC-BMR 42 and NC-BMR 90 were selected for their varying degrees of resistance to* P. tabacina. *A chemically mutated resistant line was also included. N. tabacum L. cv. KY 14 was selected as the susceptible variety. Plants were inoculated at the 4 to 5 leaf stage with a suspension of* P. tabacina. *Foliar samples were taken at 0, 2, 4, and 7 days following challenge.*

The work of Sela-Buurlage et al (1993) demonstrated that only specific isozymes of chitinases and β-1,3-glucanases have antifungal activity, and that these enzymes have a synergistic effect. Although previous studies clearly suggest a role for PR proteins in tobacco bred for resistance, their function is yet unclear. Transgenic plants overexpressing the PR-1a gene were more resistant to *P. tabacina*, however, expression of this protein does not explain the full protection observed in systemically protected plants (Alexander et al, 1993).

We have been studying the expression of chitinases and β-1,3-glucanases in tobacco lines bred for resistance to *P. tabacina*. Resistance to blue mold is considered to be due to relatively few genes acting in an additive fashion (Rufty, 1989). Several breeding lines have been developed by the use of intraspecific hybridization of wild *Nicotiana* species to *N. tabacum*. Results from SDS-PAGE and Western Blot analysis consistently revealed an earlier induction of chitinase and β-1,3-glucanase isozymes in the resistant lines. The susceptible line did not show strong expression of either chitinase or β-1,3-glucanase isozymes until after symptom development (Figure 1). Enzyme activity assays closely correlated with the Western blot analysis (Robertson and Tuzun, 1994).

Tomato

Induced resistance in tomato (*Lycopersicon esculentum* Mill) has not been as extensively characterized as in tobacco, however some interesting studies have been reported. Tigchelaar and Dick (1975) demonstrated that induced resistance to verticillium-wilt occurred in *Fusarium*-resistant tomato cultivars after dip inoculation of roots in a mixed culture of Race 1 of *Fusarium oxysporum* f. sp. *lycopersici* and *Verticillium albo-atrum*. Infection of lower leaves of tomato plants with *P. infestans* followed by a period unfavorable to disease development increased the general resistance of the plants against the pathogen (Heller and Gessler, 1986).

Studies demonstrated accumulation of PR proteins upon pre-inoculation with *P. infestans* or treatment with abiotic elicitors which induces resistance to the same organism (Christ and Mosinger, 1989). In this study a number of substances known for their ability to induce PR proteins and/or disease resistance such as, salicylic acid, ethephon, kinetin, IAA, ABA and UV treatment were tested for their ability to induce resistance to tomato diseases. Except for IAA treatment, all treatments induced the accumulation of PR proteins. Because of the antifungal nature of various PR proteins, the correlation between induced resistance and PR protein accumulation could be taken as evidence for a causal relationship between the accumulation of these proteins and the mechanisms of resistance.

Enkerli et al (1993) further demonstrated that systemic resistance to *P. infestans* in tomato was induced by infecting lower leaves with the same fungus. The level of resistance increased over a period of seven days from induction and stayed at an elevated level for at least an additional seven days. Changes in PR protein accumulation and chitinase activity were found to be induced systemically and increased in parallel with systemic resistance. β-1,3-Glucanase was found to be induced only locally. Interestingly, four varieties of tomato with different basal levels of resistance were found to develop resistance to the same level. PR protein fractions from induced plants extracted at different time intervals were tested for their antifungal activity against *P. infestans*. As expected, the antifungal activity of these fractions was found to increase in parallel with induced systemic resistance. Plants pretreated with β-aminobutyric acid were protected up to 11 days to an extent of 92% against subsequent challenge with *P. infestans* (Cohen et al, 1994). The β-isomer of aminobutyric acid also induced the accumulation of high levels of three proteins: P14a, β-1,3-glucanase, and chitinase. Even when β-aminobutyric acid was applied one day post-inoculation, protection still occurred. These results also suggest that PR proteins may be involved in induced resistance.

Biochemical defense mechanisms in tomato have been studied extensively in response to a variety of pathogens with the primary focus being upon the induction and characterization of PR proteins. Eleven acid-soluble proteins with molecular weights ranging from 13 to 82 kDa have been shown to increase in leaves infected with either *Cladosporium fulvum*, *P. infestans* or *A. solani* (Joosten and de Wit, 1989; Christ and Mosinger, 1989; Lawrence, 1993). Many of these proteins have been characterized according to their enzymatic activities. Joosten and de Wit (1989) identified two acidic (26 and 27 kDa) and two basic (30 and 32 kDa) chitinase isozymes. In this study, chitinases and β-1,3-glucanases were shown to be induced two to four days earlier in incompatible interactions of *C. fulvum* and tomato than in compatible ones. Rapid accumulation of these hydrolytic enzymes at the site of penetration of an avirulent race of *C. fulvum* could play an important role in plant defense.

In our laboratory, we are investigating the interaction between *A. solani* and tomato by utilizing a highly susceptible cultivar (Piedmont) and resistant breeding lines (71B2, NC EBR-1, and NC EBR-2). We have demonstrated that chitinases were differentially regulated in the early blight resistant breeding lines during pathogenesis with *A. solani* (Lawrence, 1993). A single basic 30 kDa chitinase isozyme was found to be constitutively expressed and shown to accumulate to a significantly greater extent during incompatible interactions than in compatible ones (Figure 2). Northern blot analysis revealed that mRNA transcripts corresponding to this basic chitinase isozyme accumulated to a high level as early as one day post-inoculation reaching maximal accumulation at two days in incompatible interactions. In compatible interactions, maximal accumulation did not occur until six days post-inoculation and not to nearly the same level as in the incompatible interaction (Figure 2). This may suggest that specificity exists at the single isozyme level (Lawrence and Tuzun, 1994). Because of the antifungal activity demonstrated by many PR proteins and the rapid accumulation of specific isozymes in incompatible situations, it is not unreasonable to hypothesize that they may play a role in plant defense.

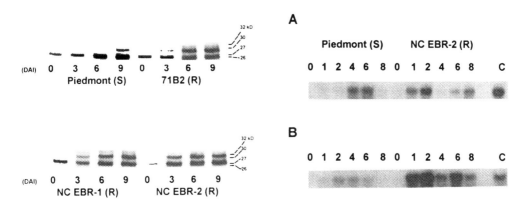

Figure 2: Western (left) and Northern blot analysis (right) of tomato chitinases. Western blot analysis was performed on Piedmont, a susceptible variety, and the three resistant lines 71B2, NC EBR-1 and NC EBR-2. Total acid soluble proteins extracted from leaf homogenates prior to and following inoculation with A. solani *(days after inoculation) were collected at different time intervals and probed with anti-tomato chitinase serum. Northern blot analysis was performed on Piedmont and NC EBR-2. The blot contains equal amounts of total RNA isolated from leaves of uninoculated plants and* A. solani *infected plants (0 to 8 days after inoculation). The same Northern blot was hybridized consecutively with labeled full-length cDNA inserts encoding the acidic tomato chitinase (panel A) and the basic tomato chitinase (panel B). Lane C is a positive control containing 15 μg of total RNA from a compatible* C. fulvum-*tomato interaction isolated 10 days after inoculation. Tomato chitinase anti-sera and acidic and basic chitinase cDNA clones were kindly provided by Prof. P. J. G. M. de Wit.*

Cabbage

Induction of hydrolytic enzymes during compatible and incompatible interactions with pathogens have been characterized in crucifers such as turnip (Dow et al, 1991) and *Arabidopsis* (Samac et al, 1990). Induction of one β-1,3-glucanase and one chitinase isozyme (CH1) is specifically associated with incompatible interactions in turnip. Another isoform (CH2) is induced only in localized areas of the inoculated leaves. In *Arabidopsis*, genes encoding a basic and acidic chitinase have been isolated and characterized. The basic chitinase gene displayed age-dependent and tissue-specific expression. This gene is induced in plants during organ specific development and by fungal infection (Samac and Shah, 1991).

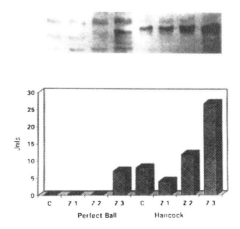

Figure 3: Western blot analysis of β-1,3-glucanases, osmotins and chitinase / lysozymes from resistant and susceptible cabbage leaves. Samples were taken from Hancock (HC, resistant) and Perfect Ball (PB, susceptible) cabbage leaves infected with bioluminescent XCC. Lanes represent H₂O-inoculated controls (C), non-bioluminescent, non-symptomatic tissue (Z1), bioluminescent, non-symptomatic tissue (Z2), and bioluminescent, symptomatic tissue (Z3). Antibodies raised against tobacco β-1,3-glucanase (BG, top) and osmotin (OS, middle) were used to detect cross-hybridizing proteins in SDS-PAGE. Similarly, antibodies raised against tobacco protein PR-P were used to detect chitinase/lysozyme cross-reactive bands (CHL, bottom). The identity of CHL-2 is indicated by an arrow at the right.

Analysis of the accumulation patterns of PR proteins in cabbage infected with *X. campestris* pv. *campestris* (Xcc), the causal organism of black rot disease, may increase our understanding of host-pathogen interactions. Our laboratory has utilized a genetically-engineered bioluminescent strain of Xcc to follow the progression of bacterial movement as well as disease progression in resistant (Hancock, HC) and susceptible (Perfect Ball, PB) cabbage varieties (Dodson et al, 1993). Acidic protein extraction and denaturing electrophoresis identified at least 12 acid extractable proteins which accumulated in both resistant and susceptible varieties following Xcc infection; however, accumulation was earlier and more pronounced in the resistant variety when compared to the susceptible variety. SDS-PAGE and Western blot analyses demonstrated the presence of a constitutively expressed chitinase/lysozyme isozyme (CHL2) in acidic extracts of HC, which was up-regulated in tissue after bacterial colonization (Fig. 3). In contrast, CHL2 accumulated in PB only after symptom development. Lysozyme activity assay results paralleled the Western blot data. Three other cabbage varieties with varying other degrees of resistance had constitutive expression of CHL2, and the level of expression was correlated with the degree of resistance.

Therefore, constitutive expression of CHL2 may be a mechanism of resistance to black rot.

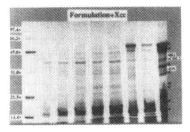

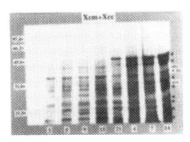

Figure 4: SDS-PAGE analysis of proteins in nonimmunized cabbage (Panel A), immunized cabbage induced with 0.2% Silwet (v / v) alone (Panel B), immunized cabbage induced with log 9.0 cfu/ml of a weakly virulent Xcc isolate B-119 (Panel C), and immunized cabbage induced with log 9.0 cfu/ml of an incompatible pathogen X. campestris pv. malvacearum (Panel D). The left side of each panel represents approximate position of proteins and molecular weights. The next five lanes show several proteins after induction. The last three lanes show an accumulation of known proteins including β-1,3-glucanase (BG) (negatively stained band, 36.3 kDa), chitinase/lysozyme 2 (CHL2) (dark band below BG, 34.6 kDa), and osmotin (OS) (26.6 kDa) as well as other unknown proteins after challenge. Molecular mass markers in kDa are given on the right.

Incompatible interactions with X. *campestris* pv. *vesicatoria* and a less pathogenic strain of Xcc induced systemic resistance in cabbage against pathogenic isolates of Xcc both under greenhouse conditions and field conditions (Jetiyanon, 1994). Inducing bacteria were infiltrated into hydathodes and stomata with the aid of an organosilicone surfactant (Silwet L-77). Silwet can develop oil-water surface tensions of greater than 30 dynes/cm and can carry bacteria under aqueous conditions directly into open stomata and/or hydathodes, their natural infection sites, without wounds or vectors. Immunized plants produced chitinase/lysozyme, β-1,3-glucanase, and osmotin as well as other pathogenesis-related proteins earlier and in greater quantities than did nonimmunized plants (Figure 4).

Conclusions

Results presented in this review indicate that constitutive expression and timely induction of specific isozymes of defense enzymes may have a role in cultivar specific resistance. This may explain why the mechanisms of resistance in breeding lines operate against only a single or a few pathogens. In contrast, in most cases we observe a generalized defense response indicated by induction of several isozymes of PR proteins as well as other proteins in systemically induced plants. This may explain the non-specific nature of resistance against multiple organisms in such plants. Since the same isozymes are induced in both cases, it appears that multiple regulatory mechanisms of expression of these genes exist. Understanding the mechanisms of multiple regulation of a single gene may result in significant advances in plant disease control strategies based on biotechnology.

We have observed, in several systems described here, a low level of constitutive expression of defense enzymes in resistant breeding lines, which may aid in initial restriction of the pathogenic organisms. More importantly, in these plants we see a boost of activity at early stages of pathogenesis. It appears that the initial activity allows plants to respond faster to the pathogenic development. There are several reports indicating that the cell wall fragments released from degrading pathogenic organisms such as glucans (Schmidt and Ebel, 1987; Cheong and Hahn, 1991) and chitin oligomers (Walker-Simons and Ryan, 1984) act as signals for plant defense responses. Initial constitutively expressed enzyme activity, though it may be in rather low levels, may be enough to release sufficient chemical signals from pathogens to induce boost of activity specifically in the infected cells and surrounding ones resulting in hypersensitive cell death.

In this review, we have concentrated on chitinases and other PR proteins. The generalized defense mechanisms, however, are more complex than can be explained by a few gene products. For example, lignification and induction of anionic peroxidases have been known to be a part of defense reactions in plants (Hammerschmidt and Kuc´, 1984). In conjuction with the CHL-2 studies, we have also been investigating the accumulation patterns of peroxidase isozymes in resistant and susceptible varieties of cabbage associated with Xcc infection. The most anionic peroxidase (pI 3.6) accumulates in both resistant and susceptible varieties with the onset of symptoms, however, accumulation is greater in the resistant varieties (Gay et al, 1994). Lagrimini et al (1993) in tobacco and Hammerschmidt and Kuc´ (1984) in cucumber have previously reported the accumulation of the most anionic peroxidases during development of resistance in these respective systems. Further studies are underway to determine what role peroxidases may play in disease defense in the cabbage-Xcc host-pathogen system.

In order to determine the role of PR proteins several strategies have been employed such as antisense expression and overexpression in transgenic plants. It has been shown that the overexpression of the PR-1a protein in transgenic tobacco resulted in a lower level of infection by two oomycete pathogens, *P. tabacina* and *P. parasitica* var. *nicotianae* (Alexander et al, 1993). Broglie et al, 1991 showed that transgenic tobacco had an increased ability to survive in soil infested with *Rhizoctonia solani*, when overexpressing a bean chitinase gene. Also, disease symptoms were delayed in development. Unfortunately, the transgenic approach has not always been wholly successful. Lagrimini et al (1993) over-expressed peroxidase in tomato and found no increase in disease resistance.

Our results indicate that the rapidity of the host response to an invading pathogen is very important in a successful defense. Furthermore, our and other results support the hypothesis that different isozymes of the PR proteins including β-1,3-glucanases and chitinases are differentially regulated and are associated with disease resistance at the physiological level. Molecular techniques and recombinant DNA methodologies, which have already led to significant advances in our understanding of plant-pathogen interactions and the mechanisms associated with disease resistance, will enable us to understand the role of specific isozymes in disease resistance. These molecular techniques will aid in the development of broad-spectrum resistance in plants without sacrificing quality and yield, a goal which has eluded plant breeders for centuries.

References

Alexander, D., Goodman, R.M., Gut-Rella, M., Glascock, C., Weymann, K., Friedrich, L., Maddox, D., Ahl-Goy, P., Luntz, T., Ward, E. & Ryals, J. (1993) *Proc. Natl. Acad. Sci.* **90**, 7327-7331.

Alstrom, S. (1991) *J. Gen. Appl. Microbiol.* **36**, 495-501.

Bent, A.F., Kunkel, B.N., Dahlbeck, D., Brown, K.L., Schmidt, R., Giraudat, J., Leung, J. & Staskawicz, B.J. (1994) *Science* **265**, 1856-1860.

Bol, J.F., Linthorst, H.J.M. & Cornelissen, B.J.C. (1990) *Annu. Rev. Phytopathol.* **28**, 113-138.

Boller, T. (1985) In: *Cellular and Molecular Biology of Plant Stress.* Key, J. & Kosuge, T. (eds.) Alan R. Liss, Inc. New York, NY. pp.247-262.

Bowles, D.J. (1990) *Annu. Rev. Biochem.* **59**, 873-907.

Broglie, K., Chet, I., Holliday, M., Cressman, R., Biddle, P., Knowlton, S., Mauvais, J. & Broglie, R. (1991) *Science* **254**, 1194-1197.

Cheong, J-J. & Hahn, M.G. (1991) *Plant Physiol.* **97**, 693-698.

Christ, U. & Mosinger, E. (1989) *Physiol. Mol. Plant Pathol.* **35**, 53-65.

Cohen, Y., Niderman, T., Mosinger, E. & Fluhr, R. (1994) *Plant Physiol.* **104**, 59-66.

Debener, T., Lahnackers, H., Arnold, M. & Dangl, J.L. (1991) *Plant J.* **1**, 289-302.

Dixon, R.A. (1986) *Biol. Rev. Camb. Philos. Soc.* **61**, 239-292.

Dodson, K.M., Shaw, J.J. & Tuzun, S. (1993) *Phytopathology* **83**, (Abst.) 1335.

Dong, X., Mindrinos, M., Davis, K.R. & Ausubel, F.M. (1991) *Plant Cell* **3**, 61-72.

Dow, J.M., Collinge, D., Milligan, D.E., Parra, R., Conrads-Strauch, J. & Daniels, M.J. (1991). In: *Biochemistry and Molecular Biology of Plant-Pathogen Interactions.* Smith, C.J. (ed.) Clarendon Press, Oxford, England. pp.163-176

Enkerli, J., Gisi, U. & Mosinger, E. (1993) *Physiol. Mol. Plant Pathol.* **43**, 161-171.

Gaffney, T., Friedrich, L., Vernooij, B., Negrotto, D., Nye, G., Uknes, S., Ward, E., Kessmann, H. & Ryals, J. (1993) *Science* **261**, 754-756.

Gay, P.A., Dodson, K.M., Robertson, T.L., Lawrence, C.B. & Tuzun, S. (1994) *Phytopathology* **84**, (Abst) in press.

Gianinazzi, S., Martin, C. & Vallee, J.C. (1970) *C. R. Acad. Sci. Paris* **270**, 2383-2386.

Heller, W.E. & Gessler, C. (1986) *J. Phytopath.* **116**, (Abst.) 323-328.

Hammerschmidt, R. & Kuc´, J. (1984) *Physiol. Plant Pathol.* **20**, 61-71.

Hennis, Y. & Chet, I. (1975) *Adv. Appl. Microbiol.* **19**, 85-111.

Hynes, R.K. & Lazarovits, G. (1989) *Can. J. Plant Pathol.* **11**, (Abst.) 191.

Jetiyanon, K. (1994) M. S. Thesis, Auburn University, Auburn, Alabama U. S. A.

Jones, J., Ashfield, T., Balint-Kurti, P., Brading, P., Dixon, M., Hammond-Kosack, K., Harrison, K., Hatzixanthis, I., Jones, D., Norcott, K. & Thomas, C. (1994) (Abstract) *Fourth Internatl. Congress of Plant Mol. Biol.*, Amsterdam, The Netherlands.

Joosten, M.H.A.J. & de Wit, P.J.G.M. (1989) *Plant Physiol.* **89**, 945-951.

Kauffman, S., Legrand, M., Geoffroy, P. & Fritig, B. (1987) *EMBO J.* **6**, 3209-3212.

Kauss, H. (1987) *Annu. Rev. Plant Physiol.* **38**, 47-72.

Keen, N.T. & Buzzell, R.I. (1991) *Theor. Appl. Genet.* **81**, 133-138.

Kloepper, J.W., Tuzun, S. & Kuc´, J. (1992) *Biocon. Sci. Technol.* **2**, 347-349.

Kobayashi, D.Y., Tamaki, S.J. & Keen, N.T. (1989) *Proc. Natl. Acad. Sci. USA* **86**, 157-161.

Kuc´, J. (1985) In: *Cellular and Molecular Biology of Plant Stress*. Key, L.& Kosuge, T. (eds.) Alan R. Liss, New York.

Kuc´, J. (1987) In: *Innovative Approaches to Plant Disease Control*. Chet, I. (ed.), John Wiley & Sons, New York.

Lagrimini, L.M., Vaughn, J., Erb, W.A. & Miller, S.A. (1993) *HortScience*. **28**, 218-221.

Lawrence, C. B. (1993) M. S. Thesis, Auburn University, Auburn, Alabama USA.

Lawrence C. B. & Tuzun S. (1994) *Phytopathology* **84**, (Abst.) .

Lawton, K., Ward, E., Payne, G., Moyer, M. & Ryals, J. (1992) *Plant Mol. Biol.* **19**, 735-743.

Linthorst, H.J. (1991) *Critical Review in Plant Sciences* **10**, 123-150.

Linthorst, H.J.M., Danash, N., Brederode, F.T., van Kan, J.A.L., de Wit, P.J.G.M. & Bol, J.F. (1991) *Mol. Plant-Microbe Interact.* **4**, 586-592.

Linthorst, H.J.M., van Loon, L.C., van Rossum, C.M.A., Mayer, A., Bol, J.F., van Roekel, J.S.C., Muelenhoff, E.J.S. & Cornelissen, B.J.C. (1990) *Mol. Plant-Microbe Interact.* **3**, 252-258.

Malamy, J., Hennig, J. & Klessig, D.F. (1992) *Plant Cell* **4**, 359-366.

Martin, G.B., Brommonschenkel, S.H., Chunwongse, J., Frary, A., Ganal, M. W., Spivey, R., Wu, T., Earle, E.D. & Tanksley, S.D. (1993) *Science* **262**, 1432-1436.

Maurhofer, M., Hase, C., Meuwly, P., Metraux, J.P. & Defago, G. (1994) *Phytopathology* **84**, 139-146.

McIntyre, J.L., Dodds, J.A. & Hare, J.D. (1981) *Phytopathology* **71**, 297-301.

Meshi, T., Motoyoshi, F., Maeda, T., Yoshiwoka, S., Watanabe, H. & Okada, Y. (1989) *Plant Cell* **1**, 515-522.

Mills, P.R. & Wood, R.K.S. (1984) *Phytopathol. Z.* **111**, 209-216.

Mindrinos, M., Katagiri, F., Yu, G.-L. & Ausubel, F.M. (1994) *Cell* **78**, 1089-1099.

Pan, S.Q., Ye, X.S. & Kuc´, J. (1992) *Phytopathology* **82**, 119-123.

Robertson, T.L. & Tuzun, S. (1994) *Phytopathology* **84** (Abst.).

Rufty, R.C. (1989) In: *Blue Mold of Tobacco*. McKeen, W.E. (ed.) The American Phytopathological Society, St. Paul. p288.

Samac, D.A., Hironaka, C.M., Yallaly, P.E. & Shah, D.M. (1990) *Plant Physiol.* **93**, 907-914.

Samac, D.A.& Shah, D.M. (1991) *Plant Cell* **3**, 1063-1072.

Schlumbaum, A., Mauch, F., Vogeli, U. & Boller, T. (1986) *Nature* **324**, 365-367.

Schmidt, E. & Ebel, J. (1987) *Proc. Natl. Acad. Sci.* USA **84**, 4117-4121.

Sela-Buurlage, M.B., Ponstein, A.S., Bres-Vloemans, S.A., Melchers, L.S., van den Elzen, P.J. M. & Cornelissen, B.J.C. (1993) *Plant Physiology* **101**, 857-863.

Shinshi, H., Neuhaus, J., Ryals, J. & Meins, F. (1990) *Plant Mol. Biol.* **14**,357-368.

Slusarenko, A.J., Croft, K.P. & Voisey, C.R. (1991) In: *Biochemistry and Molecular Biology of Host-Pathogen Interactions.* Smith, C.J. (ed.) Oxford University Press.

Stolle, K., Zook, M., Hebard, F. & Kuc´, J. (1988) *Phytopathology* **78**, 1193-1197.

Tigchelaar, E.C. & Dick, J.B. (1975) *HortScience* **10**, 623-624.

Tuzun, S. & Kloepper, J.W. (1994) In: *Proceedings of the Third International Workshop on Plant Growth-Promoting Rhizobacteria.* Ryder, M.H., Stephens, P.M. & Bowen, G.D. (eds.) CSIRO, Adelaide, Australia. pp104-109.

Tuzun, S. & Kuc´, J. (1991) In: *The Biological Control of Plant Diseases.* Book series no. 42 from the Food and Fertilizer Technology Center.

Tuzun, S. & Kuc´, J. (1989) In: *Blue Mold of Tobacco.* McKeen, W.C. (ed.) The American Phytopathological Society, St. Paul p288.

Tuzun, S. & Kuc´, J. (1985) *Phytopathology* **75**, 1127-1129.

Tuzun, S., Rao, N., Vogeli, U., Schardl, C.L. & Kuc´, J. (1989) *Phytopathology* **79**, 979-983.

van Loon, L.C. & van Kammen, A. (1970) *Virology* **40**, 199-211.

van Loon, L.C. (1985) *Plant Mol. Biol.* **4**, 111-116.

van Peer, R., Niemann, G. J. & Schippers, B. (1991) *Phytopathology* **81**, 728-734.

Vance, C.P., Kirk, T.K. & Sherwood, R.T. (1980) *Annu. Rev. Phytopath.* **18**, 259-288.

Walker-Simons, M. & Ryan, C.A.(1984) *Plant Physiol.* **76**, 787-790.

Wei, G., Kloepper, J.W. & Tuzun, S. (1991) *Phytopathology* **81**, 1508-1512.

Wei, G., Tuzun, S. & Kloepper, J.W. (1992) *Phytopathology* **82**, (Abst.) 1109.

White, R.F. (1979) *Virology* **99**, 410-412.

Whitham, S., Dinesh-Kumar, S.P., Choi, D., Hehl, R., Corr, C. & Baker, B., (1994) The product of the tobacco mosaic virus resistance gene *N*: Similarity to Toll and the interleukin-1 receptor. *Cell* **78**, 1101-1115.

Yalpani, N., Shulaev, V. & Raskin, I. (1993) *Phytopathology* **83**, 702-708.

Ye, X.S., Pan, S.Q. & Kuc´, J. (1989) *Physiol. Mol. Plant Pathol.* **35**, 161-175.

The Role of Antifungal Metabolites in Biological Control of Plant Disease

Steven Hill, Philip E. Hammer and James Ligon

Agricultural Biotechnology, Ciba-Geigy Corporation, P.O. Box 12257, Research Triangle Park, NC 27709-2257, USA

Introduction

It has long been recognized that the monoculture of wheat leads to the development of soils that are suppressive to take-all, a disease of wheat caused by *Gaeumannomyces graminis var. tritici* (Weller, 1988). This suppressiveness is due to the development of populations of microbes, especially fluorescent pseudomonads, in the soil that are inhibitory to the growth of pathogens (Cook and Weller, 1986). This has led to the discovery and study of many soil microbes that are capable of inducing suppressiveness in soils to a number of soil-borne pathogens in various crop systems (Dowling and O'Gara, 1994; Lam and Gaffney, 1993). It is known that different biological mechanisms account for this phenomenon, known as biocontrol. These mechanisms include the production of inhibitory compounds such as antibiotics (Loper and Buyer, 1991; O'Sullivan and O'Gara, 1992; Voisard et al, 1989; Keel et al, 1992; Thomashow and Weller, 1988; Howell and Stipanovic, 1980; Hill et al, 1994), the production of siderophores that limit the availability of iron to the pathogen (Kloepper et al, 1980; Becker and Cook, 1988) and direct competition between the microbial antagonist and the pathogen for nutrients and/or occupation of environmental niches (Foster, 1986; Paulitz, 1990). In many cases, biocontrol activity from a single microbe is derived from a combination of these mechanisms. In this review, we will focus primarily on the role of antifungal compounds which are increasingly believed to be the dominant factor determining biocontrol activity in many systems.

Antibiotic metabolites for which important roles in biocontrol have clearly been demonstrated include phenazine-1-carboxylate (Phz; Thomashow and Weller, 1988) and 2,4-diacetylphloroglucinol (Phl; Keel et al, 1992). Phenazine-1-carboxylate is produced by *P. fluorescens* and *P. aureofaciens* strains which

0-8493-8265-3/97/$0.00+$.50

suppress take-all disease of wheat, caused by *Gaeumannomyces graminis* var. *tritici*. The compound is active at <1mg/ml *in vitro* and its production in the rhizosphere has been demonstrated. Furthermore, the presence of the antibiotic on roots is directly correlated with disease suppression (Thomashow et al, 1990). Mutants of a phenazine-producing strain that did not produce this compound demonstrated no *in vitro* inhibition of *G. graminis* and had reduced biocontrol activity. A gene region was isolated from the parent that complemented the phenazine deficient phenotype of the mutants and these revertants were restored in their biocontrol activity. In addition, the gene region was introduced into heterologous *Pseudomonas* strains that normally did not produce Phz and did not demonstrate biocontrol activity. The exconjugants produced Phz and had biocontrol activity (Thomashow, 1991; Thomashow and Pierson, 1991; Thomashow and Weller, 1988).

The production of 2,4-diacetylphloroglucinol by *P. fluorescens* strain CHAO has been convincingly demonstrated to have an important role in biocontrol of black root rot of tobacco caused by *Thielaviopsis basicola* (Keel et al, 1992) and damping-off of sugar beet by *Pythium ultimum* (Fenton et al, 1992). Purified Phl has broad antifungal, antibacterial and phytotoxic effects, and its presence in the rhizosphere of plants inoculated with Phl producing strains has been correlated with disease suppression (Keel et al, 1992). Compared to their wild-type *Pseudomonas* parents, mutant strains deficient in Phl production were significantly less suppressive to black root rot of tobacco (Haas et al, 1991; Keel et al, 1992), take-all of wheat (Haas et al, 1991; Keel et al, 1992; Vincent et al, 1991), and *Pythium* damping-off of sugar beet (Fenton et al, 1992).

Two other groups have reported similar results in which gene regions were isolated from Phl producing wild-type strains and these regions complemented Phl⁻ mutants of the parent. The regions also transformed heterologous nonproducing strains into Phl producing strains (Fenton et al, 1992; Vincent et al, 1991). However, in only one of these cases did the introduction of the ability to produce Phl to the heterologous *Pseudomonas* strain result in significantly increased biocontrol activity (Vincent et al, 1991). While other mechanisms play a role in biocontrol activity, it is clear that Phl is an important mechanism for the suppression of fungal pathogens.

Pyrrolnitrin (Prn), a chlorinated phenylpyrrole antibiotic, is another antifungal compound that is produced by numerous *Pseudomonas* biocontrol strains and has broad antifungal activity (Arima et al, 1964; Homma et al, 1989, 1991). Purified Prn has been used to control fungal diseases of plants (Goulart et al, 1992; Hammer et al, 1993; Homma et al, 1989; Janisiewicz et al, 1991; Takeda et al, 1990), as have synthetic analogs of Prn (Gehmann et al, 1990; Nevill et al, 1988). However, research on the role of Prn in biocontrol has produced some conflicting results. Production of Prn by *Pseudomonas* strains is correlated with biocontrol effectiveness (Janisiewicz and Roitman, 1988; Homma et al, 1989; Howell and

Stipanovic, 1979). It is generally thought that Prn is an important mechanism for the biocontrol of fungal pathogens. Recent genetic studies have produced differing conclusions on the role of Prn in some biocontrol strains. McLoughlin et al (1992) generated NTG mutants of *P. cepacia* which were deficient in Prn production and lost the ability to inhibit the growth of *Sclerotinia sclerotiorum in vitro*. In growth chamber experiments, however, the mutants were as effective as the parent strain in the control of *Sclerotinia* wilt on sunflower seedlings. Similarly, Kraus and Loper (1992) studied Tn5 mutants of *P. fluorescens* strain Pf-5 which had lost the ability to produce Prn, pyoluteorin and a third uncharacterized antibiotic 3. While these antibiotic production mutants no longer inhibited *Pythium ultimum in vitro*, the biocontrol activities of the mutants against damping-off induced by this pathogen in cucumber were not affected. However, Prn is not particularly active against *Pythium* (Homma et al, 1989), so it is not possible to make conclusions about the specific role of Prn in biocontrol from this report.

In a later paper the same group described studies with another Tn5 mutant strain of Pf-5 that was specifically deficient in Prn production but still produced pyoluteorin and antibiotic 3 (Pfender et al, 1993). This mutant, in contrast to the parent strain, did not inhibit the mycelial growth of *Pyrenophora tritici* in agar culture, nor did it suppress ascocarp formation by the fungus on infested wheat straw. Mutants which were restored to Prn production by introducing a complementing gene region behaved like the wild-type parent in these respects.

Support for a role of Prn in biocontrol also comes from the group of Jayaswal et al (1992) who produced a Tn5-induced mutant of *P. cepacia* that did not synthesize Prn and had no *in vitro* antifungal activity. This mutant did not control *Diplodia maydis* corn seedling disease, a disease that was controlled by the pyrrolnitrin producing wild-type strain. Similarly, Hill et al (1994) isolated an NTG mutant of a P. *fluorescens* strain that, unlike the parent, did not produce Prn, did not inhibit *Rhizoctonia solani in vitro* and did not suppress *R. solani*-induced damping off of cotton (Figure 1). A gene region was cloned from the wild-type strain that complemented the mutant for Prn production and control of *R. solani in vitro* and in biocontrol assays. In addition, the gene region was transferred to heterologous *Pseudomonas* strains which were not previously known to produce Prn and the recombinant strains produced Prn and provided good biocontrol activity in greenhouse tests.

In a subsequent report, this group demonstrated that the lesion in the *P. fluorescens* strain is in a regulatory gene that coordinately regulates the production of chitinase, cyanide and a gelatinase in addition to Prn (Gaffney et al, 1994). A similar global regulatory gene, *gacA*, which controls the synthesis of several antifungal activities in the *P. fluorescens* strain CHAO was described by Laville et al (1992).

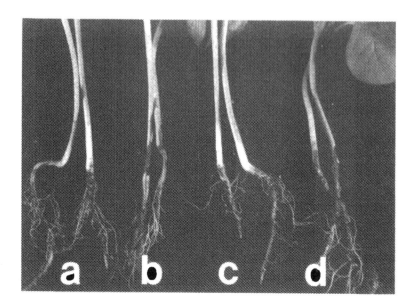

Figure 1: Biocontrol of Rhizoctonia solani-*induced damping off of cotton seedlings by wild type* Ps. fluorescens *BL915 and a BL915 orf5⁻ mutant. Cotton seedlings were sown in sterile potting medium, inoculated with* R. solani *and drenched with suspension of* 10^9 *bacteria/ml.*
a - R. solani *inoculated control; b - noninoculated control; c - inoculated, drenched with BL915; d - inoculated, drenched with BL915 orf5⁻. orf5 (a homolog of* gacA) *is a regulatory gene controlling multiple antifungal activities.*

The mutant reported by Hill et al (1994) did not produce any of the antifungal compounds produced by the parent strain and so the phenotypic defects were not limited to Prn production. Conversion of heterologous strains to Prn production and effective biocontrol activity by the introduction of the DNA region containing the *gacA* homolog was also not limited to Prn since these were later shown to produce chitinase and cyanide, in addition to Prn (Gaffney et al, 1994). This indicates that the heterologous *Pseudomonas* strains in this study had the genetic potential to produce Prn, chitinase and cyanide but the expression of these activities was repressed under the conditions used in the study. The introduction of a global regulatory gene (a transcriptional activator) from a different bacterium resulted in expression of the genes. Although there have been some conflicting reports, we believe that the evidence available clearly indicates that the production of Prn does have an important role in the biocontrol of pathogens sensitive to this compound.

The relatively recent discovery of global regulatory systems that coordinately regulate the production of a number of antimicrobial activities in biocontrol strains (Laville, et al, 1992; Gaffney et al, 1994) has significant implications for strategies whose goal is the isolation of mutants or genes confined to a single activity. Indeed, most biocontrol strains reported to date for which detailed studies are available, produce multiple compounds with antimicrobial activity. In addition, in most cases there is good evidence for coordinate regulation of these activities (Fig. 2). Therefore, care must be used in the interpretation of studies where this information is lacking for a biocontrol strain.

The comparison of a mutant that does not produce a specific antifungal compound with the parent and with genetically complemented derivatives for the purpose of analyzing the role of that compound in biocontrol is not valid unless it is known that the mutation is specific for one phenotype and is not pleotrophic. Similarly, screening strategies whose goal is the discovery of genes specific for one antifungal phenotype should account for the presence of multiple, coordinately regulated activities. A screen for mutants that is based on the loss of *in vitro* activity should use organisms that are selectively sensitive to the activity of interest and not to other activities that may be produced. Otherwise, the screen will most likely isolate regulatory mutants that are deficient in all antifungal activities.

Disease resistance in plants induced by rhizobacteria: role of metabolites

In addition to controlling soil-borne fungal pathogens, recent reports have demonstrated that some rhizosphere bacteria are able to induce resistance to invasion by foliar plant pathogens. Wei et al (1991) demonstrated that several strains of rhizobacteria, when applied to the soil, can induce resistance in cucumber to infection by *Colletotrichum orbiculare*. Bacterization of carnation seeds with *Pseudomonas* strain WCS417r induced resistance to Fusarium wilt (van Peer et al, 1991) and bacterization of bean seeds with *Pseudomonas* strain S97 induced resistance to halo blight (Alstrom, 1991). In all of theses studies the inoculated bacterial strain could not be detected on the plant leaf surfaces or in the tissues of the aerial portions of the plants. This spatial separation between the bacteria in the rhizosphere and the site of resistance in the leaves suggests that resistance was induced systemically.

Maurhofer et al (1994) recently reported the first experiments to examine the mechanisms of this type of resistance. They demonstrated that the well-characterized *P. fluorescens* strain CHAO induced resistance to tobacco necrosis virus in tobacco. The development of resistance was concomitant with the appearance of typical indicators of systemic acquired resistance (SAR) including the synthesis of pathogenesis related proteins and an increase in the concentration of salicylic acid. Salicylic acid has been shown to modulate the SAR response in plants (Gaffney et al, 1993). Maurhofer et al (1994) demonstrated that neither the production of antifungal compounds or salicylic acid by the bacterial strain affected the induction of SAR, but they presented evidence implicating the production of pyoverdine by the strain in the induction of resistance. Clearly this new and interesting area concerning the interactions between rhizobacteria and plants deserves further attention.

Regulation of antibiotic production

Restrictive growth conditions, nutrient limitation or entry into the stationary growth phase typically induce the production of antibiotics and other secondary metabolites. Studies of gene regulation have been used to dissect many of the intricacies of regulation of antibiotic production. In *Streptomyces* secondary metabolite regulation has been studied in great detail with particular emphasis on global regulatory control mechanisms (Horinouchi et al, 1990; Horinouchi and Beppu, 1992; Hong et al, 1991; Ishizuka et al, 1992). Hong et al (1991) described a two-component regulatory system consisting of a sensor kinase and a transcription activator (*afsR*) from *S. coelicolor* A3 which, when introduced into *S. lividans* caused a marked increase in the production of the pigmented antibiotics actinorhodin and undexylprodigiosin. This transcriptional activator must be phosphorylated by a phosphokinase (Afsk) in order to stimulate transcription of genes for antibiotic production (Ishizuka et al, 1992; Figure 2).

Similar two component regulatory systems are found in *Pseudomonas* spp. which regulate the production of antibiotics and secondary metabolites. Laville et al (1992) isolated a pleiotropic mutant of *P. fluorescens* strain CHAO that does not produce Phl, hydrogen cyanide or pyoluteorin. These *gacA* mutants have drastically reduced ability to suppress black root rot of tobacco, caused by *Thielaviopsis basicola*. Sacherer et al (1994) demonstrated that the *gacA* gene also is essential for the expression of two extracellular enzymes in CHAO, phospholipase C and a 47 kDa metalloprotease, both of which are normally produced in stationary phase cultures. Hrabak and Willis (1993) reported that the production of both an extracellular protease and the toxin syringomycin by *P. syringae* pv. *syringae* is regulated by *lemA* which is thought to encode the transmembrane sensor kinase protein of a two-component regulatory system.

Gaffney et al (1994) have identified a transcriptional activator in *P. fluorescens* strain BL915 that regulates the production of Prn, chitinase, cyanide and a gelatinase. This gene region, designated *orf5*, is nearly identical to *gacA*, differing in only two of 213 residues in the predicted amino acid sequences. The *orf5* region, when introduced into a pleiotropic mutant deficient in the above antifungal compounds, restores the mutant to the wild-type phenotype. Furthermore, Hill et al (1994) demonstrated that *P. fluorescens* strains BL914 and BL922, normally poor biocontrol agents, suppressed *R. solani*-induced damping off of cotton when they carried the *orf5* gene. Presumably strains BL914 and BL922 carry the structural genes needed to synthesize the antifungal compounds but did not normally produce them due to repression of gene expression under the growth conditions used. These findings shed considerable light on the regulation of secondary metabolites in bacteria. Better knowledge of the control of expression of antifungal factors may be used to develop improved biocontrol strains.

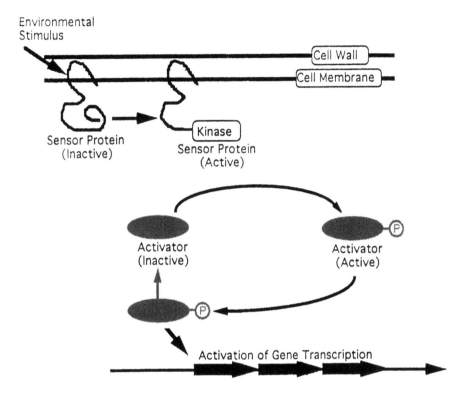

Figure 2: Depiction of a typical two component bacterial regulatory system. Upon sensing an environmental signal(s) the membrane-bound sensor/kinase protein is activated and phosphorylates a cytoplasmic transcriptional activator protein. In the phosphorylated state, this protein activates the transcription of a number of coregulated genes.

The development of biocontrol organisms that are effective in controlling plant diseases and that are commercially viable is an important goal. As pressure increases from governments around the world to reduce the use of agricultural chemicals whenever possible, biocontrol products will become more important in the future. However, in order for this to be feasible, more basic knowledge is needed in order to understand why biocontrol agents are sometimes inconsistent in their performance. This may be due to different environmental conditions that affect the regulation of the synthesis of antifungal compounds. Therefore, it is important to gain a better understanding of the factors that govern the regulation of the synthesis of these compounds in order to learn to manage this potential problem or to modify it to obtain consistently effective biocontrol. Another potential pathway to the development of improved biocontrol organisms is the creation of strains that produce elevated levels of important antifungal compounds. This could be accomplished by traditional means such as mutation and selection, or by recombinant modifications. In either case more basic information about the biochemistry and genetics of antibiotic production and its role in biocontrol is needed.

References

Alstrom, S. (1991) *J. Gen Appl. Microbiol.* **37**, 495-501.

Arima, I., Imanaka, H., Fukuta, M. & Tamura, G. (1964) *Agr. Biol. Chem.* **28**, 575-576.

Becker, J.O. & Cook, R.J. (1988) *Phytopathology* **78**, 778-782.

Cook, R.J. & Weller, D.M. (1986) In: *Innovative Approaches to Plant Disease Control.* Chet, I. ed., John Wiley & Sons. pp. 41-76.

Dowling, D.M.& O'Gara, F. (1994) *TIBTECH* 12, 133-141.

Fenton, A.M., Stephens, P.M., Crowley, J., O'Callaghan, M. & O'Gara, F. (1992) *Appl. Env. Microbiol.* **58**, 3873-3878.

Foster, R.C. (1986) *Annu. Rev. Phytopathol.* **24**, 211-234.

Gaffney, T.D., Fredrich, L., Vernooji, B., Negrotto, D., Nye, G., Uknes, S., Ward, E., Kessmann, H. & Ryals, J. (1993) *Science* 261, 754-756.

Gaffney, T.D., Lam, S.T., Ligon, J., Gates, K., Frazelle, A., DiMaio, J., Hill, S., Goodwin, S., Torkewitz, N., Allshouse, A.M., Kempf, H.-J. & Becker, J.O. (1994). *Mol. Plant-Microbe Interact.* **7**, 455-463.

Gehmann, K., Nyfeler, R., Leadbeater, A.J., Nevill, D. & Sozzi, D. (1990) In: *Proc. Brighton Crop Protection Conference - pests and diseases - 1990.* **2**, 399-406.

Goulart, B.L., Hammer, P.E., Evensen, K.B., Janisiewicz, W. & Takeda, F. (1992) *J. Amer. Soc. Hort. Sci.* **117**, 265-270.

Haas, D., Keel, C., Laville, J., Maurhofer, M., Oberhansli, T., Schnider, U., Voisard, C., Wuthrich, B. & Defago, G. (1991) In: *Advances in Molecular Genetics of Plant-Microbe Interactions.* Vol. 1. Hennecke, H. & Verma, D.P.S. (eds.). Kluwer, Dordrecht. pp 450-456.

Hammer, P.E., Evensen, K.B. & Janisiewicz, W.J. (1993) *Plant Dis.* **77**, 283-286.

Hill, D.S., Stein, J.I., Torkewitz, N.R., Morse, A.M., Howell, C.R., Pachlatko, J.P., Becker, J.O. & Ligon, J.M. (1994) *Appl. Environ. Microbiol.* **60**, 78-85.

Homma, Y., Sato, Z., Hirayama, F., Konno, K., Shirahama, H. & Suzui, T. (1989) *Soil Biol. Biochem.* **21**, 723-728.

Homma, Y., Chikuo, Y. & Ogashi, A. (1991) In: *Plant Growth-Promoting Rhizobacteria--Progress and Prospects. IOBC/WPRS Bulletin Vol.14/8.* Keel, C., Koller, B. & Defago, G. (eds.). pp 115-118.

Hong, S.-K., Kito, M., Beppu, T. & Horinouchi, S. (1991) *J. Bacteriology* **173**, 2311-2318.

Horinouchi, S., Kito, M., Nishiyaa, M., Furuya, K., Hong, S.-K., Miyake, K. & Beppu, T. (1990) *Gene* **95**, 49-56.

Horinouchi, S. & Beppu, T. (1992) *Gene* **115**, 167-172.

Howell, C.R. & Stipanovic, R.D. (1979) *Phytopathology* **69**, 480-482.

Howell, C.R. & Stipanovic, R.D. (1980) *Phytopathology* **70**, 712-715.

Hrabak, E.M. & Willis, D.K. (1993) *Mol. Plant-Microbe Interact.* **6**, 368-375.

Ishizuka, H., Horinouchi, S., Keiser, H.M., Hopwood, D.A. & Beppu, T. (1992) *J. Bacteriology* **174**, 7585-7594.

Janisiewicz W.J. & Roitman, J. (1988) *Phytopathology* **78**, 490-494.

Janisiewicz, W., Yourman, L., Roitman, J. & Mahoney, N. (1991). *Plant Disease* **75**, 490-494.

Jayaswal, R.K., Fernandez, M.A., Visintin, L. & Upadhyay, R.S. (1992) *Can. J. Microbiol.* **38**, 309-312.

Keel, C., Schnider, U., Maurhofer, M., Voisard, C., Laville, J., Burger, U., Wirthner, P., Haas, D. & Defago, G. (1992) *Mol. Plant-Microbe Interact.* **5**, 4-13.

Kloepper, J.W., Leong, J., Teintze, T. & Schroth, M.N. (1980). *Curr. Microbiol.* **4**, 317-320.

Kraus, J. & Loper, J.E. (1992). *Phytopathology* **82**, 264-271.

Lam, S.T. & Gaffney, T.D. (1993) In: *Biotechnology in Plant Disease Control.* Chet, I. (ed.) Wiley-Liss, New York. pp 291-320.

Laville, J.,Voisard, C., Keel, C., Maurhofer, M., Defago, G. & Haas, D. (1992). *Proc. Natl. Acad. Sci. USA* **89**, 1562-1566.

Loper, J.E. & Buyer, J.S. (1991) *Mol. Plant-Microbe Interact.* **4**, 5-13.

Maurhofer, M., Hase, C., Meuwly, P., Metraux, J.-P. & Defago, G. (1994) *Phytopathology* **84**, 139-146.

McLoughlin, T.J., Quinn, J.P., Betterman, A. & Bookland, R. (1992) *Appl. Environ. Microbiol.* **58**, 1760-1763.

Nevill, D., Nyfeler, R. & Sozzi, D. (1988) *Proc. Brighton crop protection conference - pests and diseases - 1988.* **1**, 65-72.

O'Sullivan, D.J. & O'Gara, F. (1992) *Microbiol. Reviews* **56**, 662-676.

Paulitz, T.C. (1990) In: *New Directions in Biological Control.* Baker, R.R. & Dunn, P.E. (eds.), Alan R. Liss, Inc., New York. pp 713-724.

Pfender, W.F., Kraus, J. & Loper, J.E. (1993) *Phytopathology* **83**, 1223-1228.

Sacherer, P., Defago, G. & Haas, D. (1994) *FEMS Microbiol. Letters* **116**, 155-160.

Takeda, F., Janisiewicz, W.J., Roitman, J., Mahoney, N. & Ables, F.B. (1990) *HortScience* **25**, 320-322.

Thomashow, L.S., Weller, D.M., Bonsall, R.F. & Pierson, L.S. (1990) *Appl. Environ. Microbiol.* **56**, 908-912.

Thomashow, L.S. (1991) In: *Plant Growth-Promoting Rhizobacteria -- Progress and Prospects.* Keel, C., Koeller, B. & Défago, G. (eds.) IOBC/WPRS Bulletin 14/8. pp 109-114.

Thomashow, L.S. & Pierson, L.S. (1991) In: *Advances in Molecular Genetics of Plant-Microbe Interactions,* Vol 1. Hennecke, H. & Verma, D.P.S. (eds.) Kluwer, Dordrecht. pp 443-449.

Thomashow, L.S. & Weller, D.M. (1988) *J. Bacteriology* **170**, 3499-3508.

van Peer, R., Nieman, G.J. & Schippers, B. (1991) *Phytopathology* **81**, 728-734.

Vincent, M.N., Harrison, L.A., Brackin, J.M., Kovacevich, P.A., Mukerji, P., Weller, D.M. & Pierson, E.A. (1991) *Appl. Env. Microbiol.* **57**, 2928-2934.

Voisard, C., Keel, C., Haas, D. & Défago, G. (1989) *EMBO J.* **8**, 351-358.

Wei, G., Kloepper, J.W. & Tuzun, S. (1991) *Phytopathology* **81**, 1508-1512.

Weller, D.M. (1988) *Annu. Rev. Phytopathol.* **26**, 379-407.

Genetically Engineered Protection of Plants against Potyviruses

Indu B. Maiti and Arthur G. Hunt

Departments of Plant Pathology and Agronomy, University of Kentucky, Lexington, KY 40546-0091, USA

Introduction

The use of virus genes as tools to generate virus resistance in plants (termed pathogen-derived resistance, or PDR) has proven to be a powerful approach for developing crops protected against viruses, and for studying aspects of virus infection in plants. Virus coat protein (CP) gene-mediated protection has been documented for a large number of plant virus families (see Fitchen and Beachy, 1993; Wilson, 1993; Beachy et al, 1990 for recent reviews). More recently, genes encoding modified and unmodified replicase proteins have been used to effect virus resistance as well (reviewed in Fitchen and Beachy, 1993; Carr and Zaitlin, 1993). These studies indicated that different virus genes may be useful agents for genetically engineering resistance to most families of plant viruses, and that several stages in the virus life cycle may be susceptible to the effects of virus-encoded proteins supplied in *trans*.

Potyviruses are among the most numerous of the different families of plant viruses (Milne, 1988). Potyviruses cause disease in most crop plants, and they can amplify the severity of disease symptoms caused by other viruses due to synergism in mixed infections (see Ross, 1968; Damirdagh and Ross, 1967 for examples). Potyviruses are single component, positive-sense RNA viruses whose genomes have proteins (genome-linked proteins, or VPgs) covalently linked to their 5' ends and polyadenylate tracts at their 3' termini (Riechmann et al (1992); see Figure 1). These genomes, which consist of some 10^4 nucleotides, encode a single polyprotein that is co- or post-translationally processed to yield at least seven and as many as eleven different virus-encoded gene products. These gene products include three proteinases that together catalyze the different proteolytic cleavages that yield the different virus-encoded proteins. In addition, most potyvirus-encoded proteins may play roles, directly or indirectly, in genome replication (Klein et al, 1994).

The preponderance of potyvirus pathogens in cultivated plants renders approaches to developing resistant varieties especially important. Because potyviruses share common genome structures and gene expression strategies, it might be expected that detailed descriptive and mechanistic information in a relatively small number of systems might lend itself to a general strategy whereby plant scientists may rapidly and efficiently engineer potyvirus protection in any plant species. To this end, a number of different approaches towards applying PDR to potyviruses have been described in the past several years. Studies with potyvirus CP genes have been most prevalent, but reports dealing with other potyvirus genes have been presented as well. This review will consider the following: what potyvirus genes have been used to generate virus-resistant transgenic plants; what is known (and not known) about the mechanisms of PDR against potyviruses; and what are the prospects for further developments in the area of PDR against potyviruses.

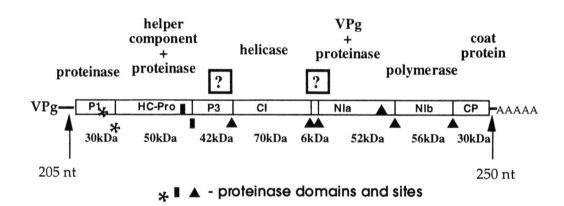

Figure 1: Genetic map of TVMV. VPg - genome-linked protein; HC - helper component; CI - cylindrical inclusion protein; NIa - VPg-proteinase; NIb - putative RNA-dependent RNA polymerase; CP - coat protein. Known and putative functions are noted above the map, and the predicted molecular weights of the respective polypeptides after proteolytic processing given below the map. Proteinase domains and their corresponding processing sites are noted as indicated below the map. Not shown are the suboptimal processing sites that have been identified in the P3 and NIa cistrons. Cistrons are not drawn in size proportion.

Potyvirus genes and strategies used to engineer resistance

A. Coat protein gene-mediated resistance against potyviruses

A number of reports have described protection against potyviruses in transgenic plants that express potyvirus coat protein (CP) genes (summarized in Table 1). In most instances, plants that express potyvirus CP genes and accumulate full sized CPs have been found to possess two defining characteristics: they exhibit a transient infection (termed recovery, and described below) when inoculated with the virus from which the CP gene was derived, and they are protected against potyviruses other than that from which the CP gene was derived. These characteristics, recovery and broad-range protection, are distinct from the properties of CP-mediated protection noted in other virus systems (Fitchen and Beachy, 1993; Beachy et al, 1990) and are hallmarks of CP-mediated protection against potyviruses.

Plants that express potyvirus CP genes typically undergo a response to infection by potyviruses that has been termed by Lindbo et al (1993) as recovery. This response includes the development of a typical infection (with occasional delays) in leaves that had been formed at the time of inoculation, followed by a gradual abatement of symptoms in newly developed leaves, and ending with leaves in which no symptoms, or virus, are apparent. Recovery has been documented in transgenic plants that express the CP genes from TEV (Lindbo and Dougherty, 1992a; see the footnotes to Table 1 for a listing of abbreviations), PPV (Ravelonandro et al, 1993), ZYMV (Fang and Grumet, 1993; Namba et al, 1992), TVMV (Maiti et al, 1993), BYMV (Hammond and Kamo, 1993) and PVY (Smith et al, 1994). In these studies, recovery required extended growth of infected plants after inoculation, an observation that may explain the absence of recovery in some reports.

The phenomenon of recovery is a general, but not universal, characteristic of transgenic plants that express potyvirus CP genes. For example, *Cucumis melo* plants that expressed the ZYMV CP gene were highly resistant to ZYMV, with most inoculated plants remaining symptom-free for the duration of the experiment, and no indications of transient infections in these experiments (Fang and Grumet, 1993). Interestingly, recovery from infections by PVY and TEV was noted in transgenic *Nicotiana tabacum* plants that expressed the ZYMV CP gene (Fang and Grumet, 1993). Whether this apparent discrepancy reflects a difference in the degree of protection in the two hosts (with recovery being essentially "instantaneous" in *Cucumis melo*) or a more fundamental difference between the different plant species remains an unanswered question.

As was the case with ZYMV in *Cucumis melo*, *N. tabacum* that expressed the PVY[N] and PVY[H] CP genes displayed apparent immunity to these respective strains of PVY (Kollar et al, 1993; Van der Vlugt et al, 1992; Farinelli and Malnoe, 1993), and to a different strain of PVY as well (Farinelli and Malnoe, 1993). Interestingly, these properties were also seen in lines that expressed "untranslatable" PVY CP RNAs (Van der Vlugt et al, 1992; Farinelli and Malnoe, 1993; Smith et al, 1994, 1995), raising the possibility that CP-mediated protection against PVY strains is an RNA-mediated phenomenon.

Potato lines that express PVY CP genes also fail to undergo the recovery so typical of potyvirus CP-containing tobacco plants. This has been reported for potato lines that express the PVY[o] gene (Lawson et al, 1990) and the PVY-A26 gene (Smith et al, 1995). In the latter study, plant lines were described in which an infection in the inoculated leaf, but not in the other leaves, induced recovery. However, most (10 of 11) protected lines that expressed a translatable CP mRNA were highly resistant (e.g., they failed to develop an infection upon challenge with PVY).

Plants that express potyvirus CP genes can be protected against potyviruses other than those from which the CP gene was derived (termed here as heterologous potyviruses) as well as the cognate homologous potyviruses. Stark and Beachy noted that plants that expressed the SMV CP gene were protected against two different potyviruses, TEV and PVY (Stark and Beachy, 1989). Similar observations were made in plants that expressed the ZYMV (Fang and Grumet, 1993; Namba et al, 1992), WMV II (Maiti et al, 1993), PPV (Ravelonandro et al, 1993), PaRV (Ling et al, 1991), and TVMV (Namba et al, 1992) CP genes. In these different studies, the degree of heterologous protection varied widely, from an incremental delay in symptom onset and fractional (not quantitative) reduction in the numbers of symptomatic plants in an experiment (Ravelonandro et al, 1993; Fang and Grumet, 1993; Namba et al, 1992; Stark and Beachy, 1989; Ling et al, 1991), to near quantitative protection, even at relatively high doses (Maiti et al, 1993). However, broad range protection is not a universal trait in plants that express potyvirus CP genes; plants that expressed the TEV (Lindbo et al, 1993), BYMV (Hammond and Kamo, 1993), and PVY (Smith et al, 1994, 1995) CP genes were not protected against other potyviruses.

One interesting aspect of the protection afforded by potyvirus CPs is the observation that significantly modified CPs can be effective in conferring protection. The 'core' ZYMV CP, which lacked the N-terminal 41 amino acids found in the native protein, was as effective in protecting against ZYMV in melon and TEV and PVY in *Nicotiana tabacum* as the full ZYMV CP (Fang and Grumet, 1993). Likewise, plants that expressed a TEV CP lacking the N-terminal 29 amino acids of this protein had characteristics similar to those of plants expressing the full TEV CP (Lindbo and Dougherty, 1992a). Addition of 14 amino acids derived from nopaline synthase had little apparent effect on the protective effects of this CP against BYMV; plants that expressed this protein had properties similar to plants that expressed BYMV CP lacking these 14 amino acids (Hammond and Kamo, 1993). These observations are consistent with studies indicating that modifications in the N-terminal domains of potyvirus CPs have little effects on biological (Klein et al, 1994) or biochemical (Jagadish et al, 1991) activities of CPs. They are also consistent with the observations that the N-terminal domains of potyvirus CPs share little by way of sequence similarity (Timmerman et al, 1990; Van der Vlugt et al, 1989).

In contrast with N-terminal modifications, deletion of the C-terminal 20 amino acids of the TEV CP had a significant change in the effects of this protein in transgenic plants (Lindbo and Dougherty, 1992a). Plants that expressed the C-terminally deleted TEV CP showed a significant delay in the development of disease, and those symptoms that appeared were milder than those seen in controls. Importantly, these plants did not recover from infection, as did plants that expressed the full-length N-terminally deleted TEV CP.

The emerging consensus from these different studies is that potyvirus CP genes indeed can be effective agents for protection against potyvirus infections, and that the characteristics of protection are largely (but not completely) predictable. However, the differences noted between several of these studies indicate that lessons learned in one system may not necessarily apply to other potyvirus-host combinations, nor to all potyviruses in a single host plant.

B. Replicase-mediated resistance

Several studies in recent years have demonstrated the utility of genes encoding RNA-dependent RNA polymerases (RDRPs) for protection against plant viruses (reviewed by Carr and Zaitlin, 1993). This utility has been shown for potyviruses as well. Tobacco plants that expressed the PVY and TVMV NIb genes (the gene encoding the RDRP) were protected against PVY or TVMV, respectively (Audy et al, 1994; Table 2). Plants that expressed the PVY NIb gene rarely developed symptoms and the resistance was effective at relatively high virus doses (Audy et al, 1994). NIb-mediated protection did not extend to other, heterologous potyviruses (Audy et al, 1994; I. B. Maiti and A. G. Hunt, unpublished observations). Interestingly, plants that expressed a PVY NIb variant, one in which the highly conserved GDD motif was altered, were not protected against PVY (Audy et al, 1994). This suggests that the PVY NIb acts at the level of protein, and that catalytically active NIb protein is required for protection.

C. Resistance mediated by the NIa gene

Potyvirus genomes encode a number of non-structural proteins as well as CPs and RDRPs. As most of these probably play roles in RNA replication (Riechmann et al, 1992; Klein et al, 1994), and because replication is a likely stage at which NIbs act to effect protection, it is possible that other potyvirus genes can confer protection in transgenic plants. This is the case with at least one NIa gene, that from TVMV. Tobacco plants that expressed the TVMV VPg/proteinase (or NIa) gene were found to be resistant to TVMV, but susceptible to TEV and PVY (Maiti

Virus[1]	CP gene(s)[2]	Plant host(s)	Viruses protected[3]	References
ZYMV	CP, "core" CP	*Cucumis melo*, *Nicotiana tabacum*	ZYMV	Ravelonandro et al, 1993
TEV	CP, CP (ΔN), CP (ΔC), CP (ΔNC), untranslatable CP	*N. tabacum*	TEV	Lindbo et al, 1993
TEV	CP, untranslatable CP	*N. tabacum*	TEV-H, -N, -S, -OX	Inokuchi & Hirashima, 1987
ZYMV	CP	*N. benthamiana*	WMV II, BYMV, PVY, PeMV, TEV PeaMV, CYVV,	Fang and Grumet, 1993
WMV II	CP	*N. benthamiana*	WMV II, BYMV, PVY, PeMV, TEV PeaMV, CYVV,	ditto
SMV	CP	*N. tabacum*	PVY, TEV	Timmerman et al, 1990
PVY[N]	CP, NIb'-CP-UTR, untranslatable CP	*N. tabacum*	PVY[N], PVY[O]	Farinelli & Malnoe, 1993
PPV	CP	*N. benthamiana*	PPV	Lindbo & Dougherty, 1992a
PPV	CP	*N. clevelandii*	PPV, PVY[N]	ditto
PPV	CP	*N. tabacum*	PVY[N]	ditto
PVY[H]	CP	*N. tabacum*	PVY[H]	Kollar et al, 1993
PVY[N]	CP, untranslatable CP	*N. tabacum*	PVY[N]	Van der Vlugt et al, 1992
TEV	CP, CP (ΔN)	*N. tabacum*	TEV	Hammond & Kamo, 1993
PaRV	CP	*N. tabacum*	TEV, PVY, PeMV	Van der Vlugt et al, 1989
TVMV	CP	*N. tabacum*	TVMV, TEV, PVY	Namba et al, 1992
BYMV	CP, CP + 14 nos aas	*N. benthamiana*	BYMV	Maiti et al, 1993
PVY[O]	CP	*Solanum tuberosum*	PVY[O]	Kaniewski, et al, 1990

Table 1: Viral coat protein gene sequences used to generate protected plants

et al, 1993). These plants rarely developed symptoms and the resistance was effective at relatively high virus doses (Maiti et al, 1993). Moreover, this protection did not extend to other heterologous potyviruses (Maiti et al, 1993). In these respects, plants that expressed the TVMV NIa gene were similar to plants that expressed potyvirus NIb genes, and strikingly different from lines that expressed the TVMV CP gene. Vardi et al (1993) also noted that a small proportion (4%) of tobacco lines that expressed derivatives of the PVY NIa gene were protected against PVY but not TMV. The significance of this report is unclear, as the PVY gene introduced into plants lacked the N-terminal 33 codons of the NIa gene, but contained the first 83 codons of the adjacent NIb gene. Nevertheless, this report demonstrates the potential utility of NIa-containing sequences in deriving protection against potyviruses in transgenic plants.

Swaney et al (1995) also described studies involving protection of tobacco against potyviruses using portions of the TEV NIa. These authors expressed different forms of the N-terminal portion of NIa (the portion corresponding to the VPg), alone and combined with the preceding 6 kD reading frame of TEV. The resulting transgenic plants possessed characteristics similar to those seen in plants that expressed translatable and untranslatable TEV CP genes. A significant number of resulting transgenic lines (between 6 and 65%, depending on the particular construction used) displayed the recovery phenomenon or complete resistance. There seemed to be little difference in the efficacy of genes capable of yielding translatable and untranslatable RNAs in this study, but genes with somewhat smaller RNA coding regions (600 nucleotides vs. 1000 nucleotides) were distinctly less effective in yielding protected plant lines. Interestingly, recovery was seen in a number of different lines, in contrast to what was noted with TVMV NIa-containing plant lines (Maiti et al, 1993).

Abbreviations for Table 1: (opposite page)

  1. *ZYMV - zucchini yellow mosaic virus; TEV - tobacco etch virus; WMV II - watermelon mosaic virus II; SMV - soybean mosaic virus; PVY - potato virus Y (different isolates are noted with superscripts); PPV - plum pox virus; PaRV - papaya ringspot virus; TVMV - tobacco vein mottling virus; BYMV - bean yellow mosaic virus.*
  2. *Coat protein derivatives introduced into plants; see the indicated references for details. Abbreviations: CP - full length CP coding region (which includes an artificial translation initiation codon); "core" CP - the conserved C-terminal core of the CP coding region; CP (DN) - an N-terminally deleted CP gene; CP (DC) - a C-terminally deleted CP gene; CP (DNC) - a CP gene with N- and C-terminal deletions; untranslatable CP-CP genes designed to yield RNAs but not CP polypeptides; NIb'- CP-UTR - the N-terminal portion of a potyvirus genome, including part of the NIb gene, all of the CP gene and 3'-untranslated region; CP + 14 nos aas-CP coding sequence to which has been added 14 codons from the nopaline synthase gene.*
  3. *Viruses for which protection was reported. Abbreviations not described in footnote 1: TEV-H, -N, -S, -OX - different isolates of tobacco etch virus; PeaMV - pea mosaic virus; CYVV - clover yellow vein virus; PeMV - pepper mottle virus.*

D. Other potyvirus genes

The utility of potyvirus CP, NIb, and NIa genes for engineering virus resistance in plants suggests that other potyvirus genes may likewise be useful for this purpose. That this may be the case is borne out by our own observations that TVMV-resistant plant lines can be obtained with all TVMV genes except the P1 gene (Table 2). Interestingly, this survey revealed two classes of TVMV genes: those that yielded resistant plant lines in a majority of cases (the HC-pro, NIa, NIb, and CP genes), and those that yielded protected lines rarely if at all (the P1, P3, and CI genes). This survey is not complete; we have only evaluated one P1 line, and have not tested the gene between the CI and NIa genes that encodes a 6kDa polypeptide. Nevertheless, this survey indicates that almost all potyvirus genes can be viewed as potential tools for pathogen-derived protection against potyviruses.

TVMV genes[a]	# of protected lines[b]	# of sensitive lines[b]
P1-HC*[c]	0	1
HC-Pro	8[d]	0
P3	1	8
CI	1	5
NIa	7	1
NIb	10	3
CP	3[d]	1

Table 2: Properties of transgenic lines expressing individual TVMV genes

a. *TVMV gene present in the transgenic By21 lines being analyzed. With the exception of the P3 and NIb lines, expression has been verified by Western blot analysis.*

b. *Number of lines tested that displayed <50% infection frequencies (protected lines) or >75% infection frequencies (sensitive lines) when inoculated with TVMV. Except where noted (see note d), responses to TVMV were indistinguishable from By21. In all cases, infection frequencies with By21 (the untransformed control) was 100%. For these screens, plants were grown in flats (8 plants/flat) and tested for their responses to these viruses as described by Maiti et al (1993). Each line was tested at least twice.*

c. *This construction includes approximately 1/3 of the HC coding region, and thus has the potential to be processed by the P1 protease (which provides a functional assay for protease function by immunoblotting).*

d. *In these lines, recovery from an initial infection (see text) was observed.*

E. Antisense - and sense - RNA-mediated resistance

As has been observed in other systems, the expression of antisense RNAs can be an effective strategy for engineering potyvirus resistance in transgenic plants. Using CP sequences, various degrees of antisense-mediated protection have been reported for ZYMV (Fang and Grumet, 1993), TEV (Lindbo and Dougherty, 1992 a,b), PVY (Farinelli and Malnoe, 1993), and BYMV (Hammond and Kamo, 1993, 1995). The range of protection seen in these instances was generally modest, with inoculated plants displaying a delay in symptom appearance and varying degrees of amelioration of symptom severity. However, occasional cases in which highly effective protection was obtained have been reported, as have instances in which the recovery phenomenon described above were noted (Hammond and Kamo, 1995).

Two interesting and unexplained exceptions to the above generalizations have been described. *Nicotiana tabacum* plants that expressed the ZYMV antisense gene displayed the same protection against the heterologous potyviruses PVY and TEV that was observed in CP plants. Moreover, protection against the heterologous potyviruses in both lines (antisense and CP) was manifest as a recovery similar to that observed in other virus-host systems. Likewise, plants that expressed the BYMV antisense CP gene (actually, a gene containing about half of the CP gene and the 3' untranslated region) recovered from an initial infection by BYMV, much as did CP plants (Hammond and Kamo, 1993).

Additional studies have revealed that "sense" RNAs, derived from genes designed to yield RNAs of the appropriate coding sense but containing small changes (nucleotide substitutions) designed to render the predicted RNAs unable to be translated, are also effective agents for protection against potyviruses. This has been observed with untranslatable RNAs derived from the TEV (Lindbo and Dougherty, 1992a,b; Dougherty et al, 1994; Goodwin et al, 1995) and PVY (Van der Vlugt et al, 1992; Farinelli and Malnoe, 1993; Smith et al, 1994) CP genes. In these reports, the degree of protection varied with the system, but in all cases a distinct lack of correlation between protection and (transgene-derived) RNA levels was noted; in fact, there seemed to be an inverse correlation at best, with the greatest degree of protection occurring in lines with the lowest amount of RNA.

These various studies indicate that RNA-based protection against potyviruses is possible; indeed, in the best cases, RNA-mediated protection results in properties (resistance to aphid-transmitted infections, almost complete absence of virus in inoculated and systemic leaves) superior to those noted in CP-containing plants (Lindbo and Dougherty, 1992b). However, as is the case with CP-mediated protection, broad generalizations concerning RNA-mediated protection against potyviruses are difficult to come by, a fact that underscores a lack of understanding of the mechanisms underlying RNA-based protection.

Possible mechanisms

The mechanisms by which potyvirus CP genes act to confer protection are not understood. This arises from the unexpected, and as yet unexplained inconsistencies seen in different systems. Although some generalizations (recovery, broad-range protection) can be made, there are exceptions that make definitive statements of mechanism difficult. There are probably several possible reasons for the diversity seen in different systems, and many of these may be traced to differences between viruses, hosts, and host-virus combinations. The involvement of potyvirus CPs in genomic RNA packaging (Jagadish et al, 1991), aphid-mediated transmission from infected to uninfected plants (Atreya et al, 1991), and cell-to-cell and long distance movement (Dolja et al, 1994) provide a host of steps in which *trans*-added CP might interfere. The probable involvement of RNA-mediated (see below) as opposed to protein-based mechanisms is another complicating matter, although RNA-based mechanisms can clearly be a factor in CP gene-mediated protection, it is difficult to rule out an additional involvement of coat protein in plant lines that express translatable CP genes and accumulate CP.

Dougherty and his colleagues have proposed a model to explain both the phenomenon of RNA-mediated protection against potyviruses (Lindbo et al, 1993; Smith et al, 1994; Dougherty et al, 1994; Goodwin et al, 1996). They suggest that protection is a manifestation of a host cell sequence-specific RNA degradation mechanism, and that this mechanism is triggered in cells in which the transgene expression exceeds a hypothetical threshold level. This model is based on a number of observations. Chief among these is a consistent lack of positive correlation between the levels of expression of transgenes and the degree of resistance in corresponding plant lines; in many instances it has been noted that lines that are most protected against challenge virus are those with the lowest level of transgene expression (Lindbo and Dougherty, 1992a,b; Van der Vlugt et al, 1992; Farinelli and Malnoe, 1993; Smith et al, 1994; Goodwin et al, 1996). That the "site" of action of RNA-mediated protection is a cytoplasmic, post-transcriptional one is evidenced by the observation that transgene transcription is high in lines that are highly resistant to virus and possess low steady state levels of transgene-derived RNA (Dougherty et al, 1994; Smith et al, 1994; Goodwin et al, 1996).

As noted elsewhere (Dougherty and Parks, 1995), this model is generally applicable to other potyviruses (Swaney et al, 1995) and to other plant RNA viruses. More importantly, it shares features with models proposed for certain types of post-transcriptional cosuppression of gene expression in plants (Matzke and Matzke, 1993). This congruity is consistent with the observation that plants in which the expression of a GUS gene is decreased by cosuppression are also protected against GUS-containing PVX, but not PVX constructs that lack GUS

(Mueller et al, 1995). This observation suggests that these hitherto unrelated areas may be linked by a common mechanism.

The matter of broad range protection in plants that express CP genes has not been satisfactorily explained. Different (but not all) potyvirus CPs can protect against heterologous viruses, but the extent of heterologous protection varies greatly with the CP, potyvirus, and possibly the host plant. Whether these differences reflect differences in sequence homology between potyviruses (as would be expected were broad-range protection be an RNA-based phenomenon similar to that proposed by Lindbo et al, 1993) or other, as yet undiscovered differences between potyviruses, remains to be determined.

The protective effects of the NIb genes are probably due to the enzymatic activity of their products, as alteration of the conserved GDD motif renders the PVY NIb ineffective for protection (Audy et al, 1994). This makes potyvirus RDRPs different from other virus-encoded enzymes; in other instances, defective RDRPs are at least equally effective in protecting organisms against virus (Carr and Zaitlin, 1993; Inokuchi and Hirashima, 1987). For potyvirus RDRPs, competition with cellular factors or with viral RNAs are possible mechanisms for protection. However, the possible involvement of other virus-encoded proteins should not be overlooked. Indeed, several potyvirus-encoded proteins have been implicated in the replication process (Riechmann et al, 1992; Klein et al, 1994), and it is possible that functional, *trans*-supplied NIb might disrupt the functioning of other virus-encoded factors in a replication complex.

As the properties of NIa lines are similar to NIb lines and to protected lines that contain antisense or untranslatable sense CP genes, it is difficult to distinguish between protein-based (as is probably the case for NIb-mediated protection) and RNA-based mechanisms in NIa lines. Protein-based mechanisms could conceivably involve the proteinase activity of the NIa protein or the VPg domain of NIa may interfere with RNA replication. Either of these functions, when supplied in *trans*, could alter the sequence or regulation of gene expression or genome replication in infected cells and thus lead to an unsuccessful infection.

Future prospects

From the first report documenting PDR against potyviruses, the prospects for developing transgenic crops protected against a wide range of potyviruses using a single coat protein gene have seemed at times quite promising. The variability seen with different CP genes and in different hosts makes this goal somewhat distant, but still feasible with the proper combination of CP gene, viruses, and host plant.

One attractive prospect is that of utilizing multiple potyvirus genes to combine desirable resistance properties, such as the broad range of protection conferred by a CP gene with the high level resistance of an NIa or NIb gene. Indeed, potyviruses provide a novel tool with which to easily engineer plants that express multiple proteins, namely the potyviral proteinases (especially the NIa proteinases). However, this strategy presumes that the effects of the individual genes will not be negated by the presence, in a single cell, of the protein products of multiple genes expressed from a nuclear gene. Preliminary studies (I. B. Maiti and A. G. Hunt, manuscript in preparation) suggest that this assumption is not valid. We have expressed the NIa, NIb, and CP genes of TVMV as a single, self-processing polyprotein in tobacco. Surprisingly, none of the resulting transgenic lines had the properties seen in CP transgenic lines (which for TVMV consists of protection against TEV and PVY as well as TVMV, and the recovery phenomenon described above). Moreover, only one of the twelve lines analyzed were protected against TVMV in any sense, as would be expected for lines expressing either the TVMV NIa or NIb genes. In the burley tobacco system, most transgenic lines containing single NIa or NIb genes are protected against TVMV. Thus, the observation that the lines containing the NIa-NIb-CP polyprotein gene are not protected against TVMV indicates that a simple additive engineering of virus resistance properties may not be easily accomplished.

Potyviruses have been shown to interact synergistically with other viruses to produce disease symptoms more severe than those seen in individually infected plants. The prospects of utilizing potyvirus genes to combat potyvirus infections directly, and other severe viral diseases indirectly, would seem to be good. However, this will require careful testing. This is because the synergism noted between potyviruses and other viruses may involve direct interactions between virus-encoded proteins (Vance et al, 1995). It is possible that the product of a single potyvirus gene may mimic the synergism seen in dually infected plants, either through a direct interaction with the products of the partner virus, or indirectly by inadvertently altering the life cycle of the potyvirus partner (as is seen in most potyvirus CP-containing plants). Another possible outcome is one in which the partner (non-poty) virus acts to negate the effect of the product of the potyvirus-derived transgene by direct interactions between the potyvirus gene product and those of the partner virus.

The ultimate goal of PDR is the production of plant lines that are significantly protected against potyviruses in a field setting. To this end, the high level resistance against PVY in potato and TEV in tobacco has been observed in field experiments, indicating that CP gene-mediated protection, involving translatable (Kaniewski et al, 1990; Whitty et al, 1994) and non-translatable (Whitty et al, 1994) genes, is effective in such settings. Importantly, the recovery phenomenon noted in TEV CP gene-expressing lines (Lindbo et al, 1993; Lindbo and Dougherty, 1992a) was also observed in field studies (Whitty et al, 1994). Although the expression of the PVY[o] CP gene had no effects on general agronomic performance of potato (Kaniewski et al, 1990), the expression of particular mutated TEV CP genes did have an effect on general plant characteristics (Whitty et al, 1994), a result that must be taken into account in intended applications. No reports dealing with broad range protection against potyviruses in a field setting have been presented, but this is perhaps the most important and appealing aspect of PDR against potyviruses and merits close scrutiny in field studies.

Perhaps most exciting is the prospect that an understanding of PDR that permits efficient, predictable engineering of resistance to viruses will lead to a detailed understanding of the basic cellular processes affected by viruses. In particular, with potyviruses it is likely that satisfactory explanations for the recovery phenomena seen in many CP-containing transgenic lines, for the lack of correlation between CP levels (either RNA or protein) and degree of protection against potyviruses, for the ability of potyvirus CP genes to confer protection against heterologous potyviruses, for sense (and antisense) RNA-mediated resistance to potyviruses, and for the abilities of other potyvirus genes to serve as protective agents will bring with them important and as yet unanticipated insight into basic cellular processes in plants.

Acknowledgements

We thank Carol Von Lanken for assistance with the studies described in Table II. This research was supported by grants from the USDA NRIGRP and THRI.

References

Atreya, P. L., Atreya, C. D. & Pirone, T. P. (1991) *Proc. Nat. Acad. Sci. USA* **88**, 7887-7891.

Audy, P., Palukaitis, P., Slack, S.A. & Zaitlin, M. (1994) *Mol. Plant-Microbe Interact.* **7**, 15-22.

Beachy, R.N., Loesch-Fries, S. & Tumer, N. (1990) *Annu. Rev. Phytopathol.* **28**, 451-474.

Carr, J.P. & Zaitlin, M. (1993) *Seminars in Virology* **4**, 339-347.

Damirdagh, I.S. & Ross, A.F. (1967) *Virology* **31**, 269-307.

Dolja, V.V., Haldeman, R., Robertson, N.L., Dougherty, W.G. & Carrington, J.C. (1994) *EMBO J.* **13**, 1482-1491.

Dougherty, W.G. & Parks, T.D. (1995) *Current Opinions in Cell Biology* **7**, 399-405.

Dougherty, W.G., Lindbo, J.A., Smith, H.A., Parks, T.D., Swaney, S. & Proebsting, W.M. (1994) *Mol. Plant Microbe Interact.* **7**, 544-552.

Fang, G. & Grumet, R. (1993) *Mol. Plant-Microbe Interact.* **6**, 358-367.

Farinelli, L. & Malnoe, P. (1993) *Mol. Plant-Microbe Interact.* **6**, 284-292.

Fitchen, J.H. & Beachy, R.N. (1993) *Annu. Rev. Microbiol.* **47**, 739-763.

Goodwin, J., Chapman, K., Swaney, S., Parks, T.D., Wernsman, E.A. & Dougherty, W.G. (1996) *Plant Cell* **8**, 95-105.

Hammond, J. & Kamo, K.K. (1993) *Acta Horticulturae* **33**, 171-178.

Hammond, J. & Kamo, K.K. (1995) *Mol. Plant Microbe Interact.* **8**, 674-682.

Inokuchi, Y. & Hirashima, A. (1987) *J. Virol.* **61**, 3946-3949.

Jagadish, M.N., Ward, C.W., Gough, K.H., Tulloch, P.A., Whittaker, L.A. & Shukla, D. D. (1991) *J. Gen. Virol.* **72**, 1543-1550.

Kaniewski, W., Lawson, C., Sammons, B., Haley, L., Hart, J., Delannay, X. & Tumer, N. (1990) *Bio/Technology* **8**, 750-754.

Klein, P., Cerezo-Rodriguez, E., Klein, R., Hunt, A.G. & Shaw, J.G. (1994) *Virology*, **204**, 759-769.

Kollar, A., Thole, V., Dalmay, T., Salamon, P. & Balazs, E. (1993) *Biochimie* **75**, 623-629.

Lawson, C., Kaniewski, W., Haley, L., Rozman, R., Newell, C., Sanders, P. & Tumer, N. (1990) *Bio/Technology* **8**, 127-134.

Lindbo, J. A. & Dougherty, W.G. (1992b)*Virology* **189**, 725-733.

Lindbo, J.A. & Dougherty, W.G. (1992a) *Mol. Plant-Microbe Interact.* **5**, 144-153.

Lindbo, J.A., Silva-Rosales, L., Proebsting, W.M. & Dougherty, W.G. (1993) *Plant Cell* **5**, 1749-1759.

Ling, K., Namba, S., Gonsalves, C., Slightom, J.L. & Gonsalves, D. (1991) *Bio/Technology* **9**, 752-758.

Maiti, I.B., Murphy, J. F., Shaw, J.G. & Hunt, A.G. (1993) *Proc. Nat. Acad. Sci. USA* **90**, 6110-6114.

Matzke, M. & Matzke, A.J.M. (1993) *Annu. Rev. Plant Physiol. Plant Mol. Biol.* **44**, 53-76.

Milne, R.G. (1988) In: *The Plant Viruses*. Vol. 4. *The Filamentous Plant Viruses*. Milne, R.G. (ed.), Plenum Press, New York, NY. pp331-335.

Mueller, E., Gilbert, J., Davenport, G., Brigneti, G. & Baulcombe, D.C. (1995) *Plant J.* **7**, 1001-1013.

Namba, S., Ling, K., Gonsalves, C., Slightom, J.L. & Gonsalves, D. (1992) *Phytopathol.* **82**, 940-946.

Ravelonandro, M., Monsion, M., Delbos, R. & Dunez, J. (1993) *Plant Sci.* **91**, 157-169.

Riechmann, J.L., Lain, S. & Garcia, J.A. (1992) *J. Gen. Virol.* **73**, 1-16.

Ross, J.P. (1968) *Plant Dis. Rep.* **52**, 344-348.

Smith, H.A., Swaney, S., Parks, T.D., Wernsman, E.A. & Dougherty, W.G. (1994) *Plant Cell* **6**, 1441-1453.

Smith, H.A., Powers, H., Swaney, S., Brown, C. & Dougherty, W.G. (1995) *Phytopathol.* **85**, 864-870.

Stark, D.M. & Beachy, R.N. (1989) *Bio/Technology* **7**, 1257-1262.

Swaney, S., Powers, H., Goodwin, J., Siva Rosales, L. & Dougherty, W.G. (1995) *Mol. Plant Microbe Interact.* **8**, 1004-1011.

Timmerman, G.M., Calder, V.L. & Bolger, L.E. (1990) *J. Gen. Virol.* **71**, 1869-1872.

Vance, V.B., Berger, P.H., Carrington, J.C., Hunt, A.G. & Shi, X.M. (1995) *Virology* **206**, 583-590.

Van der Vlugt, R.A.A., Ruiter, R.K. & Goldbach, R. (1992) *Plant Mol. Biol.* **20**, 631-639.

Van der Vlugt, R., Allefs, S., De Haan, P. & Goldbach, R. (1989) *J. Gen. Virol.* **70**, 229-233.

Vardi, E., Sela, I., Edelbaum, O., Livneh, O., Kuznetsova, L. & Stram, Y. (1993) *Proc. Nat. Acad. Sci. USA* **90**, 7513-7517.

Whitty, E.B., Hill, R. A., Christie, R., Young, J.B., Lindbo, J.L. & Dougherty, W.G. (1994) *Tobacco Sci.* **121**, 30-34.

Wilson, T.M.A. (1993) *Proc. Nat. Acad. Sci. USA* **90**, 3134-3141.

Negative Selection Markers for Plants

Mihály Czakó, Allan R. Wenck and László Márton

Department of Biological Sciences, University of South Carolina, Columbia, SC 29208, USA

Introduction

Genetic approaches towards understanding biological processes involve the use of genes, which upon expression cause immediate or conditional suppression of growth or eventual lethality. Such genes can be used as negative selection markers for eliminating a particular class of cells within an individual organism or a population.

Negative selection markers are drawn from a pool of genes that can adversely affect the cells by interfering with normal growth and development. Intracellularly expressed cytotoxins are immediate candidates for this purpose. Expression of foreign genes, that are not toxic to cells from which they are derived, overexpression of an organism's own gene, and antisense RNA can also be used to modify plant phenotype. Negative selection may function at the protein level through production of a toxic protein or a toxic metabolite upon addition of substrate (lethal synthesis), at the transcription level through antisense or cosuppression effect, or even at the DNA level through an as yet undefined mechanism. The scope of the present review is limited to those genes whose effects are deleterious to the cell or the offspring, and/or allow negative selection.

The first group of marker genes used in negative selection experiments were derived from the *Agrobacterium* T-DNA. Recently, a variety of chimeric genes that express bacterial or viral coding sequences under the control of plant promoters have been developed as negative selection markers. These genes can be categorized in terms of where and when effects are manifested and the severity of these effects. Potent cytotoxins act at the cellular level and will kill the cells. Other gene products only slow down growth, thus causing a growth disadvantage for the cells that express them. When expressed in every cell, they may be deleterious to the whole plant (organism level). Other gene products cause only the offspring to fail by inhibiting one or more processes of sexual reproduction, embryo development, or seed germination. It is possible to direct gene expression to specific tissues or organs, and/or to time expression by using tissue-specific, developmentally regulated, or inducible promoters. The intracellular target affected may be unique to certain cell types, or present only at particular times, or stages of development. For some potentially deleterious genes the target has to be artificially introduced simultaneously or in advance. Depending on the mechanism, effects may be localized to the cell where the gene is expressed (i.e., cell-autonomous action) or spread to neighboring cells. In some cases a second gene which can neutralize or counteract the deleterious one, and restore the cells to normal is available. Before a negative selection gene is chosen for a particular application, several factors must be considered including the potency, level and target of action, the temporal or spatial controllability, transient expression before chromosomal integration, the conditional nature, the availability of restorer genes, and the cell-autonomous property of a given gene.

Negative selection markers have been widely and successfully used in organisms other than plants for eliminating a particular class of cells (i.e., programmed cell death or genetic ablation), to study gene inactivation and deletion formation. These markers can also be incorporated into positive-negative selection schemes for gene targeting mediated by homologous recombination. Negative selection markers can also be employed as reporter genes to study regulatory sequences in transgenic plants and to identify mutants defective in signaling processes.

This chapter reviews the presently available negative selection marker genes, potentially new ones, and their applications, advantages, and limitations. It is clear that improved negative selection marker genes will become available in the near future. More experience will accumulate in different plant systems, new control mechanisms will be worked out, and some will be used to change plant phenotype for agronomically important purpose.

Nonconditional negative selection genes

1. Diphtheria toxin A gene

Diphtheria toxin (DTx) is a protein that selectively inhibits protein synthesis. DTx catalyzes the NAD-dependent ADP-ribosylation of elongation factor EF-2 in all eukaryotic species tested (Pappenheimer and Gill, 1972). The mature toxin is composed of two domains: the NH_2-terminal portion ('A' chain) which carries the active sites of the toxin and the COOH-terminal portion ('B' chain), which is required for binding to specific receptors and cross-membrane transport of the toxin into the target cell, but is unnecessary for intracellular toxicity. Since DTx-A fragment alone is unable to bind to cells and internalize in the absence of the diphtheria toxin B chain (Pappenheimer, 1977), the effects of DTx-A are cell autonomous. Translation of a DTx-A encoding mRNA by a eukaryotic cell is expected to be lethal. Since DTx-A acts catalytically, a single toxin molecule is sufficient to inactivate translation and kill the expressing cell (Yamazumi et al, 1978). This property has important implications for heterologous expression studies. First it is impossible for this product to accumulate in a eukaryotic cell since the first toxin molecule(s) to be synthesized will stop subsequent translation of more toxin mRNAs.

Secondly, the amount of toxin present in an expressing cell would be extremely low. Since DTx-A protein prevents its own accumulation by virtue of its extreme intracellular toxicity (instead of a quantitative aspect of gene expression) the time and place of promoter activity are revealed. The affected cells are not able to develop further and the establishment of the sequence of events and the role of different cell-lineages and tissues in development becomes possible. For this reason *DTx-A* has been attractive for assembling chimeric toxin genes for intracellular expression. The activity of *DTx-A* gene in stable (Koltunow et al, 1990; Czakó and An, 1991; Thorsness et al, 1991) and transient transformation (Czakó and An, 1991) systems in plants has been recently demonstrated.

2. *Pseudomonas* exotoxin A gene

Pseudomonas aeruginosa exotoxin A, like DTx, inhibits protein synthesis by ADP-ribosylating protein elongation factor EF-2 (Wick et al, 1990). Of its three functional domains, III is responsible for catalytic activity. Similarly to DTx, one or few exotoxin molecules are sufficient to inactivate translation in a single cell and therefore to kill the expressing cell. Nevertheless, its potency is further increased by creating a chimeric toxin called barnase toxin, a translational fusion between barnase (see Section II/5) and exotoxin A (Prior et al, 1991).

Koning et al (1992) engineered a low mammalian toxicity derivative of *Pseudomonas aeruginosa* exotoxin A for intracellular expression in plant cells by fusing the ADP-ribosylating domain of the exotoxin gene to plant regulatory sequences. Exotoxin A efficacy was demonstrated by transient expression of the modified exotoxin gene in tobacco protoplasts: the exotoxin gene inhibited the expression of a co-electroporated β-glucuronidase gene. An exotoxin gene with an introduced frameshift also effectively inhibited β-glucuronidase expression in the transient assay. The activity was presumably a result of frameshifting during translation or initiation of translation at a codon other than AUG. When fused to *Brassica* seed storage protein napin regulatory sequences, the exotoxin gene specifically arrested embryo development in transgenic *Brassica napus* seeds concomitant with the onset of napin expression. The napin/exotoxin chimeric gene did not have the same pattern of expression in tobacco as in *B. napus*; in addition to inhibiting seed development, the transgenic tobacco plants were male-sterile. Seeds failing to germinate had a distinct but variable collapsed appearance.

Co-cultivation of *Brassica* and tobacco with *Agrobacterium* carrying a napin/exotoxin A chimeric gene on a binary vector resulted in few shoots, which were delayed in development and, in the case of *Brassica*, did not root. It was assumed that expression from the napin promoter was at low levels during regeneration, thus preventing plants with the napin/in-frame exotoxin gene from developing.

3. Ribosome inactivating proteins

Ribosome-inactivating proteins (RIPs: recently reviewed by Barbieri et al, 1994) constitute a class of potent cytotoxins. RIPs act enzymatically, and approximately ten molecules of pokeweed antiviral protein (PAP) will kill an animal cell (Mir et al, 1988). There are two structural types of plant RIPs: single chain (e.g., PAP), and dimeric containing an additional binding subunit which makes them highly toxic to mammal cells (e.g., ricin from castor bean, *Ricinus communis*). RIPs are site-specific RNA N-glycosidases that catalyze the removal of a single adenine base from a conserved loop of the 25S rRNA of plant ribosomes (Prestle et al, 1992; Iglesias et al, 1993). This lesion interferes with elongation factor binding and disrupts protein synthesis. Saporin-L1 from the leaves of *Saponaria officinalis* was reported to have additional activities; it also releases adenine residues from various RNAs, from DNA, and from poly(A), but not from ATP, adenosine or dATP (Barbieri et al, 1994).

RIPs of fungal origin have ribonucleolytic activity, and cleave a single phosphodiester bond in the same target domain as plant RIPs (Jimenez and Vazquez, 1985). With the availability of cloned fungal ribotoxin genes, such as the restrictocin gene (Lamy and Davies, 1991), incorporation into suicide cassettes under the control of conditional or tightly regulated promoters will be possible.

Ricin was one of the first cytotoxins to be used for genetic ablation in mammalian cells (Landel et al, 1988). With the development of conditionally mutant ricin forms (Moffat et al, 1992) and special genetic systems, subtle forms of negative selection are possible as reviewed (O'Kane and Moffat, 1992; Lord et al, 1994). Application of RIPS for genetic ablation in plant systems is yet to be reported.

4. Ribonucleases

Overexpression of ribonuclease genes is expected to be deleterious to the cell through hydrolysis of RNAs. Two different RNase genes have been successfully used for selective destruction of the tapetum (Mariani et al, 1990), and one for ablation of the stigmatic secretory zone (Goldman et al, 1994). The barnase gene from *Bacillus amyloliquefaciens* (Hartley, 1988) and the chemically synthesized *Aspergillus oryzae* gene encoding RNase-T1 (Quaas et al, 1988) were placed under the control of the *TA29* gene promoter. Expression of these chimeric ribonuclease genes within the anther selectively destroyed the tapetal cell layer, prevented pollen formation, and led to male sterility in tobacco and oilseed rape. Barnase seemed to be more effective in tobacco than RNase T1. More than one copy of chimeric RNase T1 were required to produce male sterile anthers, but a single copy of the barnase gene was sufficient to produce male-sterile plants in tobacco.

One copy of either the chimeric barnase or the RNase T1 gene was sufficient to produce male sterile oilseed rape plants.

5. Antisense RNA

The selective suppression of gene activity by antisense RNA in plants is an effective method to generate mutant phenotypes (Rothstein et al, 1987; Rodermel et al, 1988; Van der Krol et al, 1988; Kooter and Mol, 1993). Antisense RNA has been used extensively and proposed mechanisms of action have been discussed by Bourque (1995). When designed against a vital gene, the antisense gene with act either as a nonconditional or conditional negative selection marker. A bleaching white phenotype developed in tobacco plants expressing antisense RNA to the tomato phytoene synthase gene from an autonomously replicating geminivirus vector (Kumagai et al, 1995). Overexpression of the *Arabidopsis* ankyrin-repeat containing gene RNA also caused a chlorotic phenotype due to interference with chloroplast differentiation (Zhang et al, 1992). Antisense glutamate 1-semialdehyde aminotransferase gene expression decreased the steady-state mRNA level and chlorophyll synthesis (Hofgen et al, 1994). It is possible to find conditions, e.g., in the field, where albinism is lethal; therefore antisense RNA interfering with photosynthetic and protective pigment biosynthesis has the potential which apparently has not been exploited for negative selection yet. Antisense RNA has been used for genetic ablation of flowers (Landschütze et al, 1995), and for creating male sterility (Van der Meer, 1992; Xu et al, 1995) as discussed later.

6. Inactivation of gene expression as a consequence of specific sequence duplication

Negative selection can be mediated not only by the polypeptide product of the gene, but often by the transcribed RNA, and the DNA itself. Inactivation of gene expression due to specific sequence duplication is a phenomenon subject to intensive research and reviewing (Flavell, 1994). There are various mechanisms underlying the gene silencing phenomena. Rather than trying to give an overview of the phenomena and mechanisms involved, this review deals only with cases that have direct relevance to negative selection. Tobacco leaves overexpressing the tomato phytoene synthase gene initially turned bright orange; however, they eventually bleached white (Kumagai et al, 1995). This symptom was interpreted as a case of co-suppression or sense inhibition. Genes inducing albinism have a negative selection potential as already discussed.

7. Ribozymes

A ribozyme is an RNA found in various biological systems with endonucleolytic activity and capable of self-cleavage (for review see Edington et al, 1993). The consensus nucleotide sequence at the cleavage site was called the "hammerhead" motif (Forster and Symons, 1987) in certain ribozymes. The cleavage site can be separated from the catalytic site, and target RNA recognition can be modified by changing the nucleotide sequence at the cleavage site. Suitably modified ribozymes can thus catalyze cleavage reactions in *trans* and can be targeted to arbitrarily chosen RNA sequences as proposed by Haseloff and Gerlach (1988; "Gene Shears").

While relatively high-level ribozyme expression failed to inhibit β-glucuronidase target gene activity in transiently transformed protoplasts (Mazzolini et al, 1992), other ribozymes reduced *nptII* gene expression in tobacco (Steinecke et al, 1992) and chloramphenicol acetyltransferase (CAT) gene expression in tobacco (Perriman et al, 1993, 1995). An intrinsic antisense effect was also a factor contributing to target mRNA reduction as revealed in transient expression assays (Steinecke et al, 1992). Perriman et al (1995) utilized a tRNA delivery system for effective ribozyme delivery in order to provide the high ribozyme/substrate ratio essential for ribozyme activity. Hammerhead ribozyme sequences incorporated into a tRNA and expressed from an autonomously replicating geminivirus vector, reduced target RNA levels more than either the nonembedded ribozyme or antisense RNA. The reduction in the full-length mRNA and the presence of specific cleavage products demonstrated *in vivo* cleavage of the target RNA (Perriman et al, 1995). Catalytic antisense RNAs can be created by embedding a ribozyme catalytic domain into an antisense RNA construct (Tabler and Tsagris, 1991). A ribozyme not only can be targeted to a vital host gene but be part of a two component, positive-negative selection system (Wegener et al, 1994) in a manner similar to the *nptII/anti-nptII* gene pair described below (Xiang and Guerra, 1993).

8. Aromatic oxidases

In an attempt to create a visible marker gene system, a new negative selection marker, the *Rhodococcus* indole oxidase gene, was discovered that is highly efficient in tobacco and *Arabidopsis* (Wenck et al, 1996). This gene product (Hart et. al, 1990) converts indole to indigo in *E. coli* by a presumed oxidase/oxygenase activity and has no apparent effects on growth other than a distinct blue/black color within 48 hrs. This gene was cloned into a chloroplast leader sequence containing expression vector (Guerineau et al, 1988) and then into a binary plasmid (Hajdukiewicz et al, 1994) for *Agrobacterium*-mediated gene transfer. Tranformants were selected on kanamycin containing media and monitored for indigo production. Only a few, non-indigo producing, transformed shoots were obtained using tobacco leaf discs (85% fewer than with control plasmids). Northern analysis revealed that these few shoots had very low or no detectable levels of *Rhodococcus* mRNA. There was a 91% decrease in transformation efficiency in tobacco NT1 suspension cultures.

Perhaps the most dramatic effect was seen in experiments with *Arabidopsis*. Only three transformed shoots which could not be grown to maturity were obtained in four transformation experiments (over 400 shoots were expected). Leaves had a variegated pattern, indicative, perhaps, of chloroplast damage. The reason for the higher negative selection efficiency obtained with *Arabidopsis* may be the same as that observed with the herpes simplex virus thymidine kinase gene (see below) - less variability of gene expression using *Arabidopsis*.

One possible explanation for this negative selection is a reduction in endogenous indole pools leading to either a tryptophan or IAA deficiency; however, exogenous indole did not increase the number of transformed shoots obtained. A more likely explanation is that the enzyme produces a toxic intermediate from indole or from other aromatic compounds found in the chloroplast as precursors for aromatic amino acids. The specificity is not well characterized and the possibility of unknown activities taking place in the chloroplasts would not be unreasonable. A third possibility is an interference with the machinery for chloroplast protein import and the subsequent disturbance of essential chloroplast functions.

9. Antibodies and protein binding vital cellular components

Theoretically antibodies can be targeted against vital components of the cell, including proteins and protein regulatory factors. antibodies can be expressed in plant tissues (Benvenuto et al, 1991). Depending on the type of role the target component plays, the antibody gene will act either as a nonconditional or conditional negative selection marker. An antibody produced in plant cells was used to block the regulatory factor abscisic acid to obtain a wilty phenotypic mutation in tobacco (Artsaenko et al, 1995). The wilty trait behaved as a conditional negative selection marker, as the transgenic plants needed high humidity for survival.

Theoretically DNA-binding peptides can be targeted against promoters of vital genes to block transcription. This approach is similar to those utilizing antisense RNA or ribozymes to inhibit the expression of a target gene. *In vivo* repression by a site-specific DNA-binding protein designed against an oncogenic sequence was successful in mammalian cells (Choo et al, 1994) but a negative selection gene for plants based on this approach has not yet been reported.

10. *Ras*-related genes

The *ras*-related genes, encoding small GTP-binding proteins, have been found in a number of distinct organisms, including mammals, insects, slime molds, yeasts, and higher plants (Boguski and McCormick, 1993). Although the physiological functions of these proteins have not been fully clarified, the *ras* proteins have been implicated in basic cellular functions such as cellular proliferation and terminal differentiation. Overexpression of the budding yeast *ras2* gene in *Nicotiana plumbaginifolia* cells revealed that *ras2* acted as a negative selection gene in freshly isolated protoplasts from leaves and blocked cell proliferation in cell suspension-derived protoplasts (Hilson et al, 1990). *Ras2* was able to block transient expression of the chloramphenicol acetyltransferase reporter (*cat*) gene. It also dramatically reduced stable transformation. Furthermore, the *ras2* effect was species-specific: of the 9 species tested Solanaceous species were most affected. The damaging effect also depended on the hormones used in the growth medium.

ras2 constructs in antisense orientation within the expression cassette, as well as constructs lacking a plant promoter were less but still noticeably deleterious. Deletion analysis of the gene indicated that the highly conserved N-terminal region (67 amino acid residues) is involved in the 'suicide' effect. Thus, sense and antisense constructs can result, each in a different way, in the killing of transfected cells.

Results from *ras2* and *v-Ha-ras* were corroborated by Liu and Sanford (1993), who showed that two other *ras* genes from mammalian cells, *c-Ha-ras* and *c-Ki-ras* acted in a similar manner. The introduced *ras* sequences remained fully inhibitory regardless of which promoters (inducible or tissue-specific) or which orientations (sense or antisense) were tested. This result suggests that *ras* DNA itself is directly involved in the inhibitory effect by an unknown mechanism. Involvement of the RAS protein or *ras* mRNA cannot be excluded. Introduced *ras* genes may inhibit plant cells by inducing co-suppression of endogenous *ras* or *ras*-related genes, thereby leading to arrested cell growth.

Experiments with a *ras*-related gene of plant origin, which does have a cellular counterpart, shed some light on the mechanism of inhibition (Kamada et al, 1992). The *rgp1* gene, isolated from rice, encodes a small GTP-binding protein which is related to the product of the human proto-oncogene *ras-p21*. Transgenic tobacco plants, carrying *rgp1* either as sense or antisense constructs, showed abnormal shoot and flower development. The expression of *tgp1*, a presumed tobacco homolog of *rgp1*, was markedly reduced in antisense *rgp1* transformants, whereas it was apparently unaffected in sense *rgp1* transformants. These observations suggest that the phenotypic changes in antisense transformants may be mediated by an effect on native *tgp1* mRNA, whereas in sense transformants the changes may be induced by over-production of the gene product, RGP-p25 (Kamada et al, 1992).

Conditional negative selection genes

Negative selection may be made dependent on the administration of a given nontoxic substrate which is converted into a toxic metabolite by the product of the negative selection gene (lethal synthesis). Conditional lethal genes thus offer an added element of control over the negative selection through their dependency on externally provided substrates.

1. Plant genes

Certain plant genes can be selected for or against depending on the selective agent. Use of plant genes in a positive/negative selection scheme carries limitation imposed by the need for a mutation that may not be available in a given target plant and may also present problems arising from *trans*-inactivation (Kooter and Mol, 1993).

Well known examples for dual selectable plant genes are the nitrate reductase, alcohol dehydrogenase enzymes, and enzymes in the tryptophan biosynthetic pathway. Nitrate reductase converts chlorate to toxic chlorite. The *nia* gene, however, must be constitutively expressed (Nussaume et al, 1991) to make nitrate reductase activity less sensitive to growth conditions (e.g., in *Arabidopsis*; Cheng et al, 1991) since chlorate induces nitrate reductase mRNA but not enzyme activity (LaBrie et al, 1991). Selection against cultured cells with alcohol dehydrogenase activity is possible (Widholm and Kishinami, 1988) using allyl alcohol which is converted to the more toxic acrolein. Recently, recessive anthranylate phosphoribosyl transferase mutants were isolated in *Arabidopsis* using 5-methyl-anthranylic acid plus tryptophan. Wild-type activity converts 5-methylanthranylic acid to toxic 5-methyl-tryptophan (Last and Fink, 1988). Similarly, the two subunit enzyme tryptophan synthase can be selected against using 5-fluoroindole. The α-subunit converts this compound into the toxic fluoro-tryptophan analog (Last et al, 1991).

A member of the tomato *Pto* gene family, *Fen*, confers sensitivity to fenthion (an organophosphorous insecticide) resulting in rapid cell death (Martin et al, 1994). After transformation with the *Fen* gene under the control of the cauliflower mosaic virus (CaMV) 35S promoter, tomato plants that were normally insensitive to fenthion, rapidly developed extensive necrotic lesions upon exposure to fenthion. Two related insecticides, fensulfothion and fenitrothion, also elicited necrotic lesions in *Fen*-transformed tomato plants. The mechanism whereby fenthion causes discrete necrotic lesions on Fen-tomato leaves is unknown. *Fen* is a putative serine-threonine protein kinase and participates in a signal transduction pathway involved in resistance to certain *Pseudomonas syringae pv. tomato* strains. The *Fen*-fenthion system should be useful for studying programmed cell death responses and the involvement of this process in disease resistance. It remains to be shown whether *Fen* is able to confer fenthion sensitivity to other species and thereby be usable in negative selection schemes.

2. Oncogenes from *Agrobacterium* T-DNA

The T-DNA *onc* genes of the Ti plasmid are responsible for the hormone-independent tumor growth following *Agrobacterium tumefaciens* infection (Inzé et al, 1987; Weiler and Schröder, 1987). These *onc* genes have regulatory sequences similar to those of typical plant genes. They influence the differentiation of higher plants by encoding auxin and cytokinin biosynthesis, thereby changing the endogenous hormonal balance. The protein encoded by gene 2 (also *iaaH* or *tms2*) is the enzyme indole-3-acetamide hydrolase, which converts indole-3-acetamide (IAM) into the active auxin indole-3-acetic acid (IAA). Similarly, the synthetic analog 1-naphthaleneacetamide (NAM) is converted into 1-naphthaleneacetic acid (NAA).

Several pathways have been proposed for synthesis of the naturally occurring auxin, IAA (reviewed by Marumo, 1986). According to Sembdner et al (1980)

IAM is not an intermediate in auxin biosynthesis in plants. Normal plant cells do not contain other known compounds that can be converted into biologically active auxin by the gene 2 product (Follin et al, 1985; Inzé et al, 1987). In addition, according to a survey of 27 species (Kawaguchi et al, 1991) only rice showed a homologous enzyme activity with that of the gene 2 product. Therefore, gene 2 is not expected to interfere with endogenous auxin metabolism in plants. Transgenic plants expressing gene 2, as the only T-DNA derived gene, are able to use low concentrations of IAM or NAM as the sole auxin supplement. On the other hand, increased auxin concentrations are known to be toxic to plant cells (Caboche, 1980). Consequently, in association with high concentrations of IAM or NAM, gene 2 can be used as a negative selection marker at the plant (Budar et al. 1986) or protoplast level (Depicker et al, 1988). Together with the gene 2 probe, this provided a means to study gene inactivation (Depicker et al, 1988; Renckens et al, 1992), mutagenesis (Karlin-Neumann et al, 1991; Brusslan et al, 1993) and T-DNA stability in higher plants.

Inactivation of the negative selection gene results in escape from negative selection. The gene 2 system meets the requirements for inactivation studies, because gene 2 is a conditional, dominant negative selection gene allowing selection at the cellular level, and the *Agrobacterium*-mediated gene transfer system allows the introduction of a single copy. Negative selection inactivation was used by Renckens et al (1992) to examine whether the gene 2 selection system could provide a general mechanism to trap transposable elements in a *Petunia hybrida* line harboring an active transposable element as a model system. They described the application of negative selection on *Petunia* protoplasts and germinating seeds expressing gene 2. Analysis of the resulting variant clones, however, revealed no transposable element insertions. Addition of the *in vivo* methylation inhibitor 5-azacytidine to the medium reactivated gene 2 expression in some of the variants, demonstrating that reversible DNA methylation was involved in gene 2 silencing within this system. Gene 2 deletion was also detected.

The *aux2* gene from *Agrobacterium rhizogenes* TR-DNA has a similar function to gene 2 (White et al, 1985; Offringa et al, 1986). Use of the *aux2* gene from the agropine-type plasmid pRiA4 as a dominant negative marker has been demonstrated: transgenic tobacco plants transformed with the *aux2* gene and grown *in vitro* on a medium containing (NAM or IAM) overproduce auxin and are not able to form functional root systems (Camilleri and Jouanin, 1991) - a response similar to that observed with tobacco and *Arabidopsis* transformed with gene 2 (Budar et al, 1986; Karlin-Neumann et al, 1991). More recently, the *aux2* gene was shown to act as a conditional lethal marker gene at the plant level in *Brassica oleracea* (Béclin et al, 1993). Germination of transgenic cabbage seedlings on a medium supplemented with NAM resulted in inhibition of root formation, which under field conditions would result in death.

Another *Agrobacterium tumefaciens onc* gene, *ipt* (also *tmr* or gene 4) can also be overexpressed to toxic levels from a heat-inducible soybean heat shock (HS) promoter in tobacco (Ainley et al, 1993). This gene catalyzes the initial, rate-limiting step in cytokinin biosynthesis - addition of an *iso*-pentenyl side chain to 5'-AMP to form the potent cytokinin *iso*-pentenyl-adenosine monophosphate (Akiyoshi et al, 1984; Barry et al, 1984). At 42°C, a temperature that does not interfere with the development of normal plants, the chimeric HS promoter/*ipt* gene was strong enough to kill many individual vegetative clones of a transgenic line. This gene may be useful for selection of mutants in different kinds of signal transduction pathways and promoters. The limitation is that selection based on *ipt* can be carried out only at the whole plant level and the efficiency is less than 100%.

3. Herpes simplex virus thymidine kinase gene

Negative selection markers successfully used in other eukaryotic systems have been used as starting points for developing negative selection genes usable in plant systems during the plant regeneration step after *Agrobacterium*-mediated gene transfer. The human herpes simplex virus thymidine kinase enzyme is able to phosphorylate certain nucleoside analogs (e.g., the antiviral drug ganciclovir) that are not accepted by the cells own kinase(s) and convert them into toxic nucleotide analogs that block DNA replication in mammalian cells (St. Clair et al, 1984). As such, the *HSVtk* gene acts as a conditional lethal marker, which has proven extremely useful in the enrichment for homologous recombinants in the mammalian system (Mansour et al, 1988). *HSVtk* was also used for genetic ablation (Borrelli et al, 1988).

Conditions for efficient negative selection with the HSV thymidine kinase type 1 gene after *Agrobacterium*-mediated transfer into *Arabidopsis thaliana* have been reported by Czakó and Márton (1994). Transgenic plants were morphologically indistinguishable from wild type, and exhibited normal fertility. Ganciclovir (at 10^{-5} to 10^{-4} M) drastically reduced shoot regeneration on transgenic, HSVtk+ root explants, or callus formation on HSVtk+ leaf explants, but did not affect the wild-type cultures. A 35-fold reduction in shoot regeneration 8 days after transfer to shoot induction medium was observed.

Negative selection against *HSVtk* gene activity along with kanamycin selection was also efficient in *Agrobacterium*-mediated gene transfer experiments providing the first report of a conditional negative selection at the *Agrobacterium*-mediated transformation level (Czakó and Márton, 1994). The regeneration rate on double selective (ganciclovir+kanamycin) plates was as low as the frequency of shoots normally escaping kanamycin selection in *Arabidopsis* cultures.

Species-specific differences, regulatory sequence specific differences, and variability of the expression of chimeric *HSVtk* genes have also been observed. The same *HSVtk* construct which was an efficient negative selection marker in *Arabidopsis* root culture transformation experiments was only marginally efficient in tobacco producing only delayed shoot regeneration and morphologically different shoots (filiform) from leaf discs. Another construct, containing different regulatory sequences, however, provided high level *HSVtk* mRNA expression and high sensitivity to GAN in transgenic *Nicotiana tabacum* cultures, yet, inefficient negative selection in *Agrobacterium*-mediated leaf disc transformation (Czakó et al, 1995). Individual transgenic plants can be obtained that are highly sensitive to ganciclovir, but other individuals were frequently insensitive. Variability in expression explains the inefficient negative selection in tobacco. In *Arabidopsis*, unlike in tobacco, both *HSVtk* constructs were equally well expressed. Although *HSVtk* has its species limitations it will be a useful tool for gene targeting experiments in *Arabidopsis*, and any other species where variable gene expression does not interfere with negative selection. Furthermore, a recently produced *HSVtk/nptII* fusion gene for mammalian systems (Schwartz et al, 1991) may prove useful in plant systems as a dual selectable marker.

4. *Escherichia coli* cytosine deaminase gene

The *Escherichia coli codA* (Danielsen et al, 1992) encodes cytosine deaminase (CD), which catalyzes the deamination of cytosine to uracil, and likewise, that of derivatives such as 5-fluorocytosine (5FC) to 5-fluorouracil (5FU). By itself 5FC is not toxic to higher eukaryotic cells which lack CD activity. By contrast, transgenic cells expressing CD are sensitive to 5FC, which they converted to 5FU - a precursor to the thymidylate synthase suicide substrate 5-fluoro-dUMP. Consequently, the cells are deprived of dTTP for DNA synthesis (Brockman and Anderson, 1963). This substrate-dependent negative selection has been proven effective in mammalian cells (Mullen et al, 1992), *Arabidopsis* (Perera et al, 1993), as well as *Lotus* and tobacco (Stougaard, 1993), and holds promise for a wide range of other plant species.

Expression of *codA* in transformed callus results in cell death on 5FC. On 5FC CodA⁺ tobacco and *Lotus japonicus* seedlings stop growing at the early seedling stage, *Arabidopsis* seedlings do not germinate. Some questions concerning the exact mechanism of negative selection with *codA* in plants are discussed by Stougaard (1993), i.e., that secondary effects of fluorinated pyrimidines might contribute to the negative selection in addition to inhibition of DNA synthesis.

A very attractive property of *codA* is its dual nature, that is, it provides both substrate-dependent negative and positive selection (Stougaard, 1993). Positive selection of CodA⁺ tobacco on the pyrimidine biosynthetic inhibitor N-(phosphonoacetyl)-L-aspartate is possible, by pyrimidine salvage from external cytosine. In addition, a cytosine deaminase assay for plants is available (Stougaard, 1993). The simple enzymatic assay and the independence from growth conditions used for plant culture, make *codA* selection an attractive alternative to other metabolite-dependent negative selections. The use of an analog for the selection, however, has the potential of causing random mutations.

Selection that works at the translational level should be used when mutagenesis presents a problem.

5. Pro-herbicide activating bacterial cytochrome P450 gene

The *Streptomyces griseolus* gene encodes a herbicide-metabolizing cytochrome $P450_{SUI}$ monooxygenase. It catalyzes the dealkylation of the sulfonylurea R7402. The R7402 metabolite has a nearly 500-fold greater phytotoxicity than R7402 itself. $P450_{SUI}$ was expressed in transgenic tobacco (O'Keefe et al, 1994). Because this P450 can be reduced by plant ferredoxin *in vitro*, chloroplast targeted and non-targeted expression were compared. Only transgenic plants with chloroplast directed enzyme performed the $P450_{SUI}$-mediated N-dealkylation of R7402. Because of the greater phytotoxicity of the metabolite, plants carrying active $P450_{SUI}$ were susceptible to an R7402 treatment that was harmless to plants without $P450_{SUI}$. Phenotypically, the transgenic plants were indistinguishable from control plants. Their growth was arrested upon spraying with the pro-herbicide and serious injury was evident. R7402 inhibited transgenic seedlings: no roots developed but the cotyledons expanded. $P450_{SUI}$ expression was measured by antigen level and R7402 sensitivity in both enzymatic and phenotypic assays. $P450_{SUI}$ expression provides only marginal resistance to normally phytotoxic herbicides, but combined with R7402 pro-herbicide treatment, it can be used as a negative selection system in plants.

Suppressors for positive selection markers

Negative selection can be applied against a second, positive selection marker. Such suppressors find use in gene-targeting experiments. Positive selection markers for gene-targeting can be made dependent on the loss of a second, suppressor gene residing outside the homologous region. The transformant will survive positive selection only if the suppressor gene can not inactivate it. In this way, a single selection should recover predominantly homologous recombination events. Candidates for suppressor genes are antisense genes and ribozymes.

Antisense RNA methodology can be used in a positive-negative selection system for gene targeting and has shown to be useful in mammals (To et al, 1986) with the neomycin phosphotransferase II gene, the most widely used selectable marker in plants (Reynaerts et al, 1988). In plants, the anti-*nptII* gene construct was able to suppress *nptII* expression both transiently and in stably transformed tobacco calli (Xiang and Guerra, 1993). Calli harboring both the sense and antisense gene constructs did not grow or grew slowly on kanamycin medium. The promoter strength might partially contribute to the reduced sense *nptII* transcripts: the sense gene was expressed from the *nos* promoter, while the antisense gene was expressed from the 35S promoter, the latter being considered much stronger (Sanders et al, 1987). The antisense-sense principle is potentially applicable to all known positive selection genes provided that suppression is sufficient enough to lower the resistance of the transformant to the selective agent.

Ribozymes can be targeted to positive selection genes, and may even be more effective than an antisense RNA because of the combination of their enzymatic activity with an intrinsic antisense effect. An *nptII* specific ribozyme was introduced into tobacco transgenic for the neomycin phosphotransferase gene (Wegener et al, 1994). NptII levels were decreased in nearly one third of the double transformants. A correlation between reduced *nptII* expression and ribozyme expression was proven by crossing a transgenic plant expressing *nptII* with a plant carrying only the ribozyme gene. Both steady state levels of *nptII* mRNA and amounts of active enzyme were decreased in all progeny containing both transgenes, but no *nptII* mRNA cleavage product was reproducibly detected. The data indicated that, at least in stable transformants, a large excess of ribozyme over target was not a prerequisite for achieving a significant reduction in target gene expression. If this reduction results in significant increase in kanamycin sensitivity, the ribozyme may be used as a suppressor of the *nptII* positive selection gene. However, no data were provided on kanamycin sensitivity of suppressed plants.

Applications for analysis of biological processes

1. Genetic ablation

Understanding development in multicellular organisms requires the ability to dissect processes involved in cell-to-cell communication. Physical removal, or ablation, of certain cell types is one technique for identifying their contribution to positional information of neighboring cells, however, selective expression of a cell inhibitory gene is an effective alternative. Whereas physical methods, such as laser ablation, are very precise, they are not practical for organisms with small, inaccessible cells. By contrast, genetic methods can be used in any organism that can be genetically transformed (O'Kane and Moffat, 1992).

Genetic ablation methodology depends on the availability of tightly regulated cell and tissue-specific promoters in order to specifically target expression of the toxic gene. Genetic ablation is a powerful tool for studying the temporal and spatial patterns of gene expression, developmental processes in general, and the role specific cell types play in complex tissues. Application of chimeric cytotoxic genes to induce dominant 'missing pattern' mutations well complements the methodology that is currently used (for a review see Dawe and Freeling, 1991) for studying cell lineages, and provides information on cell interaction processes (for a review see Knox, 1992) during the course of spatially and temporally organized cell divisions, subsequent expansion and differentiation.

Cytotoxins that act catalytically, such as diphtheria toxin, *Pseudomonas* exotoxin, ribonucleases, and ribosome inactivating proteins are likely to be well suited for genetic ablation because they will disrupt normal cell function faster than proteins that affect DNA replication (e.g., thymidine kinase, Borrelli et al, 1988). This genetically targeted approach was first used in mammalian systems using diphtheria toxin (Breitman et al, 1987; Palmiter et al, 1987). Tightly regulated tissue specific promoters fused to a toxin gene are expected to allow the normal development of transgenic plants until the onset of promoter activity. It is possible to ablate just about any cell in the plant so long as the promoter is not expressed at an earlier time during development, e.g., in the regenerating callus. Genetic ablation as an experimental tool in dissecting plant developmental processes has been increasingly exploited since the first demonstration of its feasibility (Koltunow et al, 1990; Mariani et al, 1990).

Chimeric *DTx-A* genes have been used to study cell interaction processes during anther, stigma and seed development. The tapetal cell specific tomato *TA29* gene transcriptional control sequences have been fused to the *DTx-A* gene to determine whether *DTx-A* can be used as a cell-autonomous negative selection gene (Koltunow et al, 1990). Expression of the chimeric *TA29/DTx-A* gene within the anther selectively destroyed the tapetal cell layer indicating that tapetal cell specification occurred before *TA29* gene transcriptional activation. Destruction of the tapetum surrounding the pollen sac prevented pollen formation but did not affect the differentiation and/or function of other specialized cell types such as the surrounding sporophytic tissues.

The S-locus glycoprotein (*SLG*) gene promoter of *Brassica*, derived from the locus that controls the self-incompatibility behavior of the pistil and the specific recognition between pollen and stigma by encoding ribonucleases (Lee et al, 1994; Murfett et al, 1994) was used to direct expression of *DTx-A* in transgenic tobacco and *Arabidopsis*. Analysis of different transgenic hosts showed that the precise pattern of *SLG* activity is species dependent. The *SLG* promoter targeted toxic gene expression and cell death to the pistil and pollen in transgenic tobacco (Thorsness et al, 1991), dramatically altering floral development and fertility. *Arabidopsis* transformants showed distinct stigma and anther structural defects leading to self-sterility (Thorsness et al, 1993). However, the transformants were cross-fertile with untransformed plants. The viable pollen of ablated plants was rescued by wild-type stigmas, and, the ablated papillar cells allowed the growth of wild-type pollen (Thorsness et al, 1993). This was not the case with the distantly related *Brassica napus*: ablation of papillar cell function resulted in the loss of stigma receptivity to pollination (Kandasamy et al, 1993).

Stigma-specific expression of the barnase gene also resulted in female sterility in tobacco (Goldman et al, 1994). Pistils of transgenic plants carrying the barnase coding sequence fused to the tobacco stigma-specific *STIG1* gene 5' regulatory region underwent normal development, but lacked the stigmatic secretory zone. Pollen grains germinated on the ablated 'stigmatic' surface, but were unable to penetrate the transmitting tissue of the style. The expression of the barnase gene in an anther tapetum-specific manner led to male sterility (Mariani et al, 1990) as discussed above.

DTx-A along with *cat* and *GUS* reporter genes were instrumental in the characterization of the *Arabidopsis thaliana* γ_1-tubulin gene promoter (Kim and An, 1992). Transgenic tobacco plants carrying *DTx-A* under the control of the γ_1-tubulin promoter were of normal phenotype but seed production was drastically reduced. Furthermore, the transgene could not be transmitted to the next generation through pollen. These results point to the exclusive pollen-specific activity of the γ_1-tubulin promoter.

Diphtheria toxin-mediated cell ablation in developing male gametophytes of tobacco demonstrated the vegetative cell restricted activity of the tomato lat52 promoter (Twell, 1995). Due to ablation of the vegetative cell, the generative cell failed to undergo normal migration away from the pollen grain wall into the vegetative cell cytoplasm, and pollen-lethality eventually resulted.

Important information about the sequence of events and the role of different cell-lineages and tissues during seed development have been obtained by the use *DTx-A* for genetic ablation (Czakó et al, 1992). Transcription of the pea seed storage protein vicilin gene is regulated in the same temporal fashion in transgenic tobacco as it is in peas (Higgins et al, 1988) and is, therefore, a good candidate for genetic ablation experiments. A pea vicilin promoter-*DTx-A* gene behaved as a dominant, seed-lethal, Mendelian factor in *Arabidopsis* and tobacco (Czakó et al, 1992). Germination deficiency resulted from distinct developmental abnormalities. Species-specific differences in promoter expression and patterns of cell-to-cell interactions between *Arabidopsis* and tobacco have been observed. The timing and distribution of lesions followed a pattern typical for protein body accumulation in wild-type seeds, i.e., in the enlargement phase of cell differentiation.

The *vicilin/DTx-A* chimeric gene regulation was tight enough to prevent premature killing and to permit the regeneration of stably transformed plants using *Agrobacterium*, however, this construct decreased transient expression of a coelectroporated *cat* reporter gene (Czakó and Márton, unpublished), although not as much as the constitutive CaMV 35S RNA promoter/*DTx-A* chimeric gene (Czakó and An, 1991). Transient expression of the 35S/*DTx-A* chimeric gene appears to be responsible for the serious drop in transformation frequencies observed when an *Agrobacterium* strain carrying *DTx-A* linked to a kanamycin resistance marker was mixed with another strain carrying the same (Czakó and An, 1991) or a different (hygromycin) selectable marker (Czakó, Luong, and Márton, unpublished). Even when the hygromycin strain was mixed only at 1:1 ratio with the 35S/*DTxA-nptII* strain hygromycin resistant transformants were already almost unobtainable.

Cell ablation revealed that expression from the bean phaseolin promoter is rigorously confined to microsporogenesis and embryogenesis (van der Geest et al, 1995). The *DTx-A* region was placed under the control of the β-phaseolin promoter sequences. Transgenic tobacco, which was phenotypically normal until flowering, produced anthers that were externally normal but contained no viable pollen. The tapetum was normal, but the pollen mother cells did not undergo meiosis and subsequently degenerated, resulting in male sterility. Phaseolin promoter expression during tobacco microsporogenesis was confirmed by detection of β-glucuronidase expression in pollen from the same promoter. Seed storage protein gene promoter driven expression of *DTx-A* may not affect each tobacco cultivar equally. Cultivar Xanthi used by van der Geest et al (1995) and Czakó et al (1992) did show full or at least partial male sterility, while Petit Havana SR1 did not (Czakó et al, 1992). Embryo development was also affected in a very specific way. Expression of *DTx-A* defined the heart stage as the earliest time for phaseolin expression during embryogenesis (van der Geest et al, 1995). After fertilization of the male-sterile transgenic plants with pollen from wild-type tobacco, 50 % of the resulting embryos aborted at the heart stage. This type of growth arrest is in contrast with the case of tobacco embryos expressing *DTx-A* from the pea vicilin promoter and probably reflects differences in the onset of gene expression from the two promoters (Czakó et al, 1992).

Antisense RNA can also be used for genetic ablation. An anther-specific gene, *Bcp1*, was isolated from *Brassica campestris* that was expressed in the diploid tapetum and the haploid microspores (Theerakulpisut et al, 1991). Antisense RNA expressed from the *Bcp1* gene promoter, and the haploid pollen-specific *LAT52* promoter from tomato, respectively, resulted in arrested pollen development in *Arabidopsis* (Xu et al, 1995). Aborted pollen grains appeared as empty, flattened exine shells at maturity because they lacked cytoplasmic contents.

Another example of genetic ablation by antisense RNA is the specific disintegration of the potato ovary tissue by antisense repression of mitochondrial citrate synthase (Landschütze et al, 1995). The transgenic plants carrying the antisense gene were indistinguishable from the wild type during vegetative growth. However, flower buds formed late, and aborted at an early stage of development. The importance of the tricarboxylic acid cycle during the transition from vegetative to generative phase was highlighted by negative selection.

In order to validate any conclusions reached from genetic ablation experiments utilizing negative selection genes, it is often vital to know that the action of the toxic gene product is confined to the cells where it is synthesized. The cell autonomous action has, however, usually been inferred rather than proven. As the gene products are toxic at very low levels, there is an inherent problem in detecting their presence in any cell, as the cell may be killed before they accumulate to levels sufficient to allow detection by standard immunological methods. In addition, the presence of at least some apparently normal neighboring cells in the immediate vicinity of the of the cells expressing the negative selection gene is a strong argument against damage caused by uptake of toxic gene product that has leaked from dying cells expressing the negative selection gene. Therefore, the negative selection genes used for ablation differ from native toxins in that the secretory signal sequences and regions of the protein necessary for extracellular toxin to cross the membrane (e.g., the B-chains of ricin, diphtheria toxin, and *Pseudomonas* exotoxin A) are removed. The cross-membrane mobilities of other potential negative selection gene products, and nucleic acids (e.g., ribozymes) are not well known. It will be interesting to test whether the negative effects of *HSVtk* and *codA*, respectively, are cell autonomous and, therefore, they are usable in genetic ablation experiments. This application will rely on the apparent good translocation of ganciclovir and 5-FC throughout the plant but will require that phosphorylated ganciclovir and fluorinated pyrimidines like 5FU are not exported - the possibility of the latter has thus far not been excluded (Stougaard, 1993; Czakó et al, unpublished).

Another potential limitation is that ablation relatively late in a cell's development, such as that caused in papillar cells by the SLG promoter (Kandasamy et al, 1993; Thorsness et al, 1993), does not eliminate the cell entirely. The presence of the wall or other cell remnants can complicate the interpretation of experiments intended to assess the role of cell-cell communication in development. On the other hand, the very remnants of the cells ablated relatively late in development permit observation of the structures affected, e.g., an intracellularly degenerated embryo in tobacco (Czakó et al, 1992). Genetic ablation led to a novel tissue interaction in this case: when the tobacco embryo used up all the endosperm including the remnants of ablated cells it also absorbed the parenchyma layers of the integument which are normally obliterated by and incorporated into the endosperm.

With all its limitations, genetic ablation is a powerful technique, and with the development of conditionally mutant toxigenes, such as temperature-sensitive cytotoxins (Bellen et al, 1992; Moffat et al, 1992), an *amber* mutation (Kunes and Steller, 1991), as well as chemically regulated promoters (Ward et al, 1993) the ability to carry out even subtler forms of cell ablation is now in hand.

2. Positive-negative selection for gene targeting

Positive-negative selection strategies (using for example *HSVtk*) have been successfully applied to mouse embryo stem cells (Capecchi, 1989). It is this use of negative selection that makes gene targeting so fruitful in mammalian systems (Mansour et al, 1988). Negative selection has not been utilized so far in gene-targeting experiments in higher plants (Offringa et al, 1992, 1993). Gene-targeting experiments are currently restricted to model genes that permit positive selection. An efficient negative selection procedure, which discriminates against nonhomologous recombinants, is crucial for isolating rare homologous recombinants in genes without a positively selectable phenotype. Without negative selection, a large population of transformants must be generated, maintained, and screened. The positive-negative strategy is as follows: a positive selectable marker gene is placed within the genomic flanking sequences, and a negative selection gene resides outside the homologous region. If random integration or a nonhomologous recombination event occurs after transformation, the negative selection gene is incorporated into the genome and the transformant will be inviable or subject to conditional negative selection. If the negative selection marker is not incorporated and only the positive selectable gene is integrated into the genome (directed by the homologous sequences) the transformant will survive positive selection. The limitations of this approach is that the negative selection gene may be lost or inactivated in ways other than homologous recombination-mediated disjunction and exclusion from the chromosome.

3. Genetic dissection of signal transduction pathways

Negative selection genes can be employed in a directed genetic approach for dissecting signal transduction pathways. This approach involves the fusion of a regulated promoter to a negative selectable reporter gene and stable transformation of a model plant with the product. The reporter gene confers a phenotype such that mutants in a biochemically defined process, such as transcriptional response, is crippled and can survive selection. Although infrequently used, such promoter fusion approaches have been useful in higher eukaryotes as well as in yeast and bacteria.

A major factor that had limited this particular genetic approach in plants is the lack of markers shown to be capable of conferring a regulatable phenotype for negative selection. Karlin-Neumann et al (1991) demonstrated that gene 2 fused to the *Arabidopsis cab140* promoter behaved as a phytochrome-regulated conditional lethal selection marker in *Arabidopsis*. Selection for phytochrome signal transduction mutants using the chimeric gene 2 resulted in such by-products as auxin resistant mutants, and an interesting promoter mutant causing cosuppression (Brusslan et al, 1993). These results suggest that this approach may be useful for selection of mutants in many different kinds of signal transduction pathways and promoters.

Potential applications to crop improvement

1. Male sterility

Improvement of crop plants through the production of hybrid varieties is a major goal of plant breeding. Genetic engineering strategies facilitating hybrid seed production have been reviewed (Goldberg et al, 1993). Induction of male sterility was amenable to genetic engineering by negative selection markers. Dominant male sterility markers can be obtained by destruction of floral or reproductive tissues, or by causing metabolic perturbances. A chimeric barnase ribonuclease gene under the control of the tapetal cell-specific *TA29* gene promoter from tobacco (Koltunow et al, 1990) was shown to act as a nuclear male sterility gene (Mariani et al, 1990) in oilseed rape. It is often necessary to reverse the dominant male sterility trait, especially in hybrids where the fruit is to be harvested. Fertility was restored by crossing male sterile plants with male fertile plants expressing the corresponding inhibitor protein, barstar, expressed in the same cells (Mariani et al, 1992).

Overexpression of the *Agrobacterium rhizogenes rolC* gene also induced male-sterility in tobacco (Schmülling et al, 1988; Spena et al, 1992) and potato (Fladung, 1990). Antisense *rolC* RNA restored fertility in tobacco (Schmülling et al, 1993).

Antisense inhibition of flavonoid biosynthesis in *Petunia* anthers resulted in male sterility (van der Meer et al, 1992). The *Petunia* chalcone synthase gene in antisense orientation was expressed from a CaMV 35S RNA promoter amended with an "anther box" for anther-specific expression. The antisense chalcone synthase RNA was expressed in the tapetum cells, and caused white anthers. Transgenic plants with white anthers were male sterile due to an arrest in male gametophyte development.

Antisense Bcp1 RNA expressed from the *Brassica Bcp1* male fertility promoter resulted in arrested pollen development in *Arabidopsis* (Xu et al, 1995). Plants homozygous for the antisense gene were completely male sterile, while hemizygous plants were leaky. The antisense male fertility gene is potentially useful for creating sterility in related crops.

Male sterility was also created by expressing in the tapetum a bacterial cytochrome P450 that catalyzes activation of a sulfonylurea pro-herbicide (O'Keefe et al, 1994; see above).

2. Pathogen resistance strategies

Cell wall matrix localization of the potent RIP PAP (Ready et al, 1986) capable of inactivating even the 'conspecific' pokeweed (*Phytolacca americana*) ribosomes is the basis for implicating PAP in a local suicide model supposed to operate in the hypersensitive reaction of pokeweed to viral infection (Bonnes et al, 1994). It has been proposed that coentrance of any pathogen and RIP should cause cell death and thereby prevent infection of the whole plant (Ready et al, 1986; Taylor and Irvin, 1990; Kataoka et al, 1992; Taylor et al, 1994). It has also been considered that saporins, and possible other RIPs, which act on substrates other than rRNA, could directly inhibit the replication of viruses by damaging their genomic or messenger RNAs (Barbieri et al, 1994). Practical application of RIPs and other potent cell-autonomous negative selection genes in antiviral and other pathogen resistance strategies requires spatial separation of the toxin and ribosomes in cells. This separation happened naturally with preproricin expressed in tobacco leaf cells. The RIP was secreted into the apoplast (Sehnke et al, 1994) where proper cleavage of the linker peptide and production of active ricin took place, but the compartmentalization prevented the destruction of tobacco ribosomes (Sehnke et al, 1994; Tagge et al, 1996).

Single-chain RIPS were also produced in transgenic plants, e.g., α-trichosanthin from *Trichosanthes kirilowii* in *Nicotiana benthamiana* (Kumagai et al, 1993), barley RIP, PAP. Expression of barley RIP under the control of a wound- and pathogen-inducible potato *wun1* gene promoter led to increased fungal protection in transgenic tobacco plants (Logemann et al, 1992). Broad-spectrum virus resistance resulted in tobacco and potato plants expressing PAP (Lodge et al, 1993).

Inhibition of fungal disease development in plants was indeed possible by engineering controlled cell death (Strittmatter et al, 1995). The controlled generation of necrotic lesions at infection sites was analogous to the naturally occurring hypersensitive cell death. Two transgenes were introduced into tobacco: the barnase and barstar genes under the control of the pathogenesis-specific potato *prp-1* and the CaMV 35S RNA promoter, respectively. Barstar is the specific inhibitor of the barnase, introduced to counteract barnase from the potential background expression in uninfected tissues. *Phytophthera infestans* growth and reproduction was considerably reduced on leaves from transgenic potato lines carrying the two-component system.

Ziljstra and Hohn (1992) proposed to utilize the special activity of the CaMV gene VI which promotes translation from dicistronic expression units to construct CaMV resistant transgenic *Arabidopsis* plants. CaMV viral infection would allow for expression of a negative selection gene located on a dicistronic expression unit leading to elimination of infected cells and inhibiting viral spread. This process thus resembles the hypersensitive reaction mentioned above.

It is conceivable that other genetic systems can be constructed wherein an incoming pathogen triggers a hypersensitive suicide reaction mediated by deleterious gene expression.

References

Ainley, W.M., McNeil, K.J., Hill, J.W., Lingle, W.L., Simpson, R.B., Brenner, M.L., Nagao, R.T. & Key, J.L. (1993) *Plant Molec. Biol.* **22**, 13-23.

Akiyoshi, D.E., Klee, H., Amasino, R.M., Nester, E.W. & Gordon, M.P. (1984) *Proc. Natl. Acad. Sci. USA* **81**, 5994-5998.

Artsaenko, O., Peisker, M., Nieden, U. zur, Fiedler, U., Weiler, E.W., Müntz, K. & Conrad, U. (1995) *Plant J.* **8**, 745-750.

Barbieri, L., Goroni, P., Valbonesi, P., Castiglioni, P. & Stirpe, F. (1994) *Nature* **372**, 624.

Barbieri, L., Battelli, M.G. & Stirpe, F. (1993) *Biochim. Biophys. Acta* **1154**, 237-282.

Barry, G.F., Rogers, S.G., Fraley, R.T. & Brand, L. (1984) *Proc. Natl. Acad. Sci. USA* **81**, 4776-4780.

Béclin, C., Charlot, F., Botton, E., Jouanin, L. & Doré, C. (1993) *Transgen. Res.* **2**, 48-55.

Bellen, H.J., D'Evelyn, D., Harvey, M. & Elledge, S.J. (1992) *Development* **114**, 787-796.

Benvenuto, E., Ordás, R.J., Tavazza, R., Ancora, G., Biocca, S., Catteneo, A. & Galeffi, F. (1991) *Plant Mol. Biol.* **17**, 865-874.

Boguski, M.S. & McCormick, F. (1993) *Nature* **366**, 643-654.

Bonnes, M.S., Ready, M.P., Irvin, J.D. & Mabry, T.J. (1994) *Plant J.* **5**, 173-183.

Borrelli, E., Heyman, R., Hsi, M. & Evans, R.M. (1988) *Proc. Natl. Acad. Sci. USA* **85**, 7572-7576.

Bourque, J.E. (1995) *Plant Science* **105**, 125-149.

Breitman, M.L., Clapoff, S., Rossant, J., Tsui, L.C., Glode, L.M., Maxwell, I.H. & Bernstein, A. (1987) *Science* **238**, 1563-1565.

Brockman, R.W. & Anderson, E.P. (1963) In: *Metabolic Inhibitors, a Comprehensive Treatise.* Vol. 1. Hochster, R.M. & Quastel, J.H. (eds.) Academic Press, New York, pp 239-285.

Brusslan, J.A., Karlin-Neumann, G.A., Huang, L. & Tobin, E.M. (1993) *Plant Cell* **5**, 667-677.

Budar, F., Deboeck, F., Van Montagu, M. & Hernalsteens, J.-P. (1986) *Plant Sci.* **46**, 195-206.

Caboche, M. (1980) *Planta* **149**, 7-18.

Camilleri, C. & Jouanin, L. (1991) *Mol. Plant-Microbe Interact.* **4**, 155-162.

Capecchi, M.R. (1989) *Science* **244**, 1288-1292.

Cheng, C.-L., Acedo, G.N., Dewdney, J., Goodman, H.M. & Conkling, M.A. (1991) *Plant Physiol.* **96**, 275-279.

Choo, Y., Sánchez,-Garcia, I. & Klug, A. (1994) *Nature* **372**, 642-645.

Czakó, M. & An, G. (1991) *Plant Physiol.* **95**, 687-692.

Czakó, M. & Márton, L. (1994) *Plant Physiol.* **104**, 1067-1071.

Czakó, M., Jang, J.-C., Herr, J.M. Jr, & Márton, L. (1992) *Mol. Gen. Genet.* **235**, 33-40.

Czakó, M., Marathe, R.P., Xiang, C., Guerra, D.J., Bishop, G.J., Jones, J.D.G. & Márton, L. (1995) *Theor. Appl. Genetics* **91**, 1242-1247.

Danielsen, S., Kilstrup, M., Barilla, K., Jochimsen, B. & Neuhard, J. (1992) *Molec. Microbiol.* **6**, 1335-1344.

Dawe, K. & Freeling, M. (1991) *Plant J.* **1**, 3-8.

Depicker, A.G., Jacobs, A.M. & Van Montagu, M.C. (1988) *Plant Cell Rep.* **7**, 63-66.

Edington, B.V., Dixon, R.A., & Nelson, R.S. (1993) In: *Transgenic Plants, Fundamentals and Applications.* NewYork, M. Dekker. pp 301-323.

Fladung, M. (1990) *Plant Breeding* **104**, 295-304.

Flavell, R.B. (1994) *Proc. Natl. Acad. Sci. USA* **91**, 3490-3496.

Follin, A., Inzé, D., Budar, F., Genetello, C., Van Montagu, M. & Schell, J. (1985) *Mol. Gen. Genet.* **201**, 178-185.

Forster, A.C. & Symons, R.H. (1987) *Cell* **49**, 211-220.

Goldberg, R.B., Beals, T.P. & Sanders, P.M. (1993) *Plant Cell* **5**, 1217-1229.

Goldman, M.H.S., Goldberg, R.B. & Mariani, C. (1994) *EMBO J.* **13**, 2976-2984.

Guerineau, G., Woolston, S., Brooks, L. & Mullineaux, P. (1988) *Nucleic Acids Res.* **16**, 11380.

Hajdukiewicz, P., Svab, Z. & Maliga, P. (1994) *Plant Mol Biol.* **25**, 989-994.

Hart, S., Kerby, R. & Woods, D.R. (1990) *J. Gen. Microbiol.* **136**, 1357-1363.

Hartley, R.W. (1988) *J. Mol. Biol.* **202**, 913-915.

Haseloff, J. & Gerlach, W. (1988) *Nature* **334**, 585-591.

Higgins, T.J.V., Newbigin, E.J., Spencer, D., Llewellyn, D.J. & Craig, S. (1988) *Plant Mol. Biol.* **11**, 683-695.

Hilson, P., Dewulf, J., Delporte, F., Installe, P., Jacquemin, P.-M., Jacobs, M. & Negrutiu, I. (1990) *Plant Mol. Biol.* **14**, 669-685.

Hofgen, R., Axelsen, K.B., Kannangara, C.G., Schuttke, I., Pohlernz, H.D., Willmitzer, L., Grimm, B. & Wettstein, D. (1994) *Proc. Natl. Acad. Sci. USA* **91**, 1726-1730.

Iglesias, R., Arias, F.J., Rojo, M.A., Escarmis, C., Ferreras, J.M. & Girbés, T. (1993) *FEBS Lett.* **325**, 291-294.

Inzé, D., Follin, A., Van Onckelen, H., Rüdelsheim, P., Schell, J. & Van Montagu, M. (1987) In: *Molecular Biology of Plant Growth Control.* Fox, J.A. & Jacobs, M. (eds.), (UCLA Symp. Mol. Biol. 44) Alan R. Liss, New York, pp 181-196.

Jimenez, A. & Vazquez, D. (1985) *Annu. Rev. Microbiol.* **39**, 649-672.

Kamada, I., Yamaguchi, S., Youssefian, S. & Sano, H. (1992) *Plant J.* **2**, 799-807.

Kandasamy, M.K., Thorsness, M.K., Rundle, S.J., Goldberg, M.L., Nasrallah, J.B. & Nasrallah, M.E. (1993) *Plant Cell* **5**, 263-275.

Karlin-Neumann, G.A., Brusslan, J.A. & Tobin, E.M. (1991) *Plant Cell* **3**, 573-582.

Kataoka, J., Habuka, N., Miyano, M., Masuta, C. & Koiwai, A. (1992) *Plant Mol. Biol.* **20**, 1111-1119.

Kawaguchi, M., Kobayashi, M., Sakurai, A. & Syóno, K. (1991) *Plant Cell Physiol.* **32**, 143-149.

Kim, Y.-H. & An, G. (1992) *Transgen. Res.* **1**, 188-194.

Knox, J.P. (1992) *Plant J.* **2**, 137-141.

Koltunow, A.M., Truettner, J., Cox, K.H., Wallroth, M. & Goldberg, R.B. (1990) *Plant Cell* **2**, 1201-1224.

Koning, A., Jones, A., Fillatti, J.J., Comai, L. & Lassner, M.W. (1992) *Plant Mol. Biol.* **18**, 247-258.

Kooter, J.M. & Mol, J.N.M. (1993) *Curr. Opin. Biotechnol.* **4**, 166-171.

Kumagai, M.H., Turpen, T.H., Weinzettl, N., Della-Cioppa, G., Donson, J., Hilf, M.E., Grantham, G.L., Dawson, W.O. & Chow, T.P. (1993) *Proc. Natl. Acad. Sci. USA* **90**, 427-430.

Kumagai, M.H., Donson, J., Della-Cioppa, G., Harvey, D., Henley, K. & Grill, K. (1995) *Proc. Natl. Acad. Sci. USA* **92**, 1679-1683.

Kunes, S. & Steller, H. (1991) *Genes Devel.* **5**, 970-983.

LaBrie, S.T., Wilkinson, J.Q. & Crawford, N.M. (1991) *Plant Physiol.* **97**, 873-879.

Lamy, B. & Davies, J. (1991) *Nucleic Acids Res.* **19**, 1001-1006.

Landel, C.P., Zhao, J., Bok, D. & Evans, G.A. (1988) *Genes Devel.* **2**, 1168-1178.

Landschütze, V., Willmitzer, L. & Müller-Röber, B. (1995) *EMBO J.* **14**, 660-666.

Last, R.L. & Fink, G.R. (1988) *Science* **240**, 305-310.

Last, R.L., Bissinger, P.H., Mahoney, D.J., Radwanski, E.R. & Fink, G.R. (1991) *Plant Cell* **3**, 345-358.

Lee, H.-S., Huang, S. & Kao, T.-H. (1994) *Nature* **367**, 560-563.

Liu, Z.R. & Sanford, J.C. (1993) *Plant Molec. Biol.* **22**, 751-765.

Lodge, J.K., Kaniewski, W.K. & Tuner, N.E. (1993) *Proc. Natl. Acad. Sci. USA* **90**, 7089-7093.

Logemann, J., Jach, G., Tommerup, H., Mundy, J. & Schell, J. (1992) *Bio/Technology* **10**, 305-308.

Lord, M., Roberts, L.M. & Robertus, J.D. (1994) *FASEB J.* **8**, 201-208.

Mansour, S.L., Thomas, K.R. & Capecchi, M.R. (1988) *Nature* **336**, 348-352.

Mariani, C., De Beuckeleer, M., Truettner, J., Leemans, J. & Goldberg, R.B. (1990) *Nature* **347**, 737-741.

Mariani, C., Gossele, V., De Beuckeleer, M., De Block, M., Goldberg, R.B., De Greef, W. & Leemans, J. (1992) *Nature* **357**, 384-387.

Martin, G.B., Frang, A., Wu, T., Brommomschenkel, S., Chunwongse, J., Earle, E.D. & Tanksley, S.D. (1994) *Plant Cell* **6**, 1543-1552.

Marumo, S. (1986) In: *Chemistry of Plant Hormones*. Takahashi, N. (ed.), CRC Press, Boca Raton, pp 49-56.

Mazzolini, L., Axelos, M., Lescure, N. & Yot, P. (1992) *Plant Molec. Biol.* **20**, 715-731.

Mir, L.M., Banoun, H. & Paoletti, C. (1988) *Exp. Cell Res.* **175**, 15-25.

Moffat, K.G., Gould, J.H., Smith, H.K. & O'Kane, C.J. (1992) *Development* **114**, 681-687.

Mullen, C.A., Kilstrup, M. & Blaese, R.M. (1992) *Proc. Natl. Acad. Sci. USA* **89**, 9933-9937.

Murfett, J., Atherton, T.L., Mou, B., Gasser, C.S. & McClure, B.A. (1994) *Nature* **367**, 563-566.

Nussaume, L., Vincent, M. & Caboche, M. (1991) *Plant J.* **1**, 267-274.

Offringa, I.A., Melchers, L.S., Regensburg-Tuink, A.J.G., Costantino, P., Schilperoort, R.A. & Hooykaas, P.J.J. (1986) *Proc. Natl. Acad. Sci. USA* **83**, 6935-6939.

Offringa, R., van den Elzen, P.J.M. & Hooykaas, P.J.J. (1992) *Transgen. Res.* **1**, 114-123.

Offringa, R., Franke-van Dijk, M.E.I., de Groot, M.J.A., van den Elzen, P.J.M. & Hooykaas, P.J.J. (1993) *Proc. Natl. Acad. Sci. USA* **90**, 7346-7350.

O'Kane, C.L. & Moffat, K.G. (1992) *Curr. Opin. Genet. Devel.* **2**, 602-607.

O'Keefe, D.P., Tepperman, J.M., Dean, C., Leto, K.J., Erbes, D.L., & Odell, J.T. (1994) *Plant Physiology* **105**, 473-482.

Palmiter, R.D., Behringer, R.R., Quaife, C.J., Maxwell, F., Maxwell, I.H. & Brinster, R.L. (1987) *Cell* **50**, 435-443.

Pappenheimer, A.M. Jr. (1977) *Ann. Rev. Biochem.* **46**, 69-94.

Pappenheimer, A.M. Jr. & Gill, D.M. (1972) In: *Molecular Mechanisms of Antibiotic Action on Protein Synthesis and Membranes*. Munoz, E., Garcia-Ferrandiz, F. & Vazques, D. (eds). Elsevier, Amsterdam, pp 134-149.

Perera, R.J., Linard, C.G. & Signer, E.R. (1993) *Plant Mol. Biol.* **23**, 793-799.

Perriman, R., Bruening, G., Dennis, E.S. & Peacock, W.J. (1995) *Proc. Natl. Acad. Sci. USA* **92**, 6175-6179.

Perriman, R., Graf, L. & Gerlach, W.L. (1993) *Antisense Res. & Development* **3**, 253-263.

Prestle, J., Schönfelder, M., Adam, G. & Mundy, K.-W. (1992) *Nucleic Acids Res.* **20**, 3179-3182.

Prior, T.I., Fitzgerald, D.J. & Pastan, I. (1991) *Cell* **64**, 1017-1023.

Quaas, R., McKeown, Y., Stanssens, P., Frank, R., Blöcker, H. & Hahn, U. (1988) *Eur. J. Biochem.* **173**, 617-622.

Ready, M.P., Brown, D.T. & Robertus, J.D. (1986) *Proc. Natl. Acad. Sci. USA* **83**, 5053-5056.

Renckens, S., de Greve, H., Van Montagu, M. & Hernalsteens, J.-P. (1992)
Mol. Gen. Genet. **233**, 53-64.

Reynaerts, A., De Block, M., Hernalsteens, J.-P. & Van Montagu, M. (1988) *Plant Molec. Biol. Manual* **A9**, 1-16.

Rodermel, S.R., Abbott, M.S. & Bogorad, L. (1988) *Cell* **55**, 673-681.

Rothstein, S.J., DiMaio, J,. Strand, M. & Rice, D. (1987) *Proc. Natl. Acad. Sci. USA* **84**, 8439-8443.

Sanders, P.R., Winter, J.A., Barnason, A.R., Roger, S.J. & Fraley, R.T. (1987) *Nucleic Acids Res.* **15**, 1543-1558.

Schmülling, T., Schell, J. & Spena, A. (1988) *EMBO J.* **7**, 2621-2629.

Schmülling, T., Röhrig, H., Pilz, S., Walden, R. & Schell, J. (1993) *Mol. Gen. Genet.* **237**, 385-394.

Schwartz, F., Maeda, N., Smithies, O., Hickey, R., Edelmann, W., Skoultchi, A. & Kucherlapati, R. (1991) *Proc. Natl. Acad. Sci. USA* **88**, 10416-10420.

Sehnke, P.C., Pedrosa, L., Paul, A.L., Frankel, A.E. & Ferl, R.J. (1994) *J. Biol. Chem.* **269**, 22473-22476.

Sembdner, G., Gross, D., Liebisch, H.-W. & Schneider, G. (1980) In: *Encyclopedia of Plant Physiology*, New series, vol 9: MacMillan, J. (ed) Springer Verlag Berlin, pp 181-244.

Spena, A., Estruch, J.J., Prinsen, E., Nacken, W., van Onckelen, H. & Sommer, H. (1992) *Theor. Appl. Genet.* **84**, 520-527.

St. Clair, M.H., Miller, W.H., Miller, R.L., Lambe, C.U. & Furman, P.A. (1984) *Antimicrob. Agents Chemother.* **25**, 191-194.

Steinecke, P., Herget, T. & Schreier, P.H. (1992) *EMBO J.* **11**, 1525-1530.

Stougaard, J. (1993) *Plant J.* **3**, 755-761.

Strittmatter, G., Janssens, J., Opsomer, C. & Botterman, J. (1995) *Bio/Technology* **13**, 1085-1088.

Tabler, M. & Tsagris, M. (1991) *Gene* **108**, 175-183.

Tagge, E.P., Chandler, J., Harris, B., Czakó, M., Willingham, M.C., Burbage, C., Afrin, L. & Frankel, A.E. (1996) *Protein Expression and Purification* (in press).

Taylor, B.E. & Irvin, J.D. (1990) *FEBS Lett.* **273**, 144-146.

Taylor, S., Massiah, A., Lomonosoff, G., Roberts, L.M., Lord, J.M. & Hartley, M. (1994) *Plant J.* **5**, 827-835.

Theerakulpisut, P., Xu, H., Singh, M.B., Pettitt, J.M. & Knox, R.B. (1991) *Plant Cell* **3**, 1037-1048.

Thorsness, M.K., Kandasamy, M.K., Nasrallah, M.E. & Nasrallah, J.B. (1991) *Dev. Biol.* **143**, 173-184.

Thorsness, M.K., Kandasamy, M.K., Nasrallah, M.E. & Nasrallah, J.B. (1993) *Plant Cell* **5**, 253-261.

To, R.Y.-L., Booth, S.C. & Neiman, P.E. (1986) *Mol. Cell. Biol.* **6**, 4758-4762.

Twell, D. (1995) *Protoplasma* **187**, 144-154.

Van der Geest, A.H.M., Frisch, D.A., Kemp, J.D. & Hall, T.C. (1995) *Plant Physiol.* **109**, 1151-1158.

Van der Krol, A.R., Mol, J.N.M. & Stuitje, A.R. (1988) *Biotechniques* **6**, 958-976.

Van der Krol, A.R., Lenting, P.E., Veenstra, J., Van der Meer, I.M., Gerats, A.G.M., Mol, J.N.M. & Stuitje, A.R. (1988) *Nature* **333**, 866-869.

Van der Meer, I.M., Stam, M.E., Van Tunen, A.J., Mol, J.N.M. & Stuitje, A.R. (1992) *Plant Cell* **4**, 253-262.

Ward, E.R., Ryals, J.A. & Miflin, B.J. (1993) *Plant Mol. Biol.* **22**, 361-366.

Wegener, D., Steinecke, P., Herget, T., Petereit, C., Philipp, C. & Schreier, P.H. (1994) *Molecular and General Genetics* **245**, 465-470.

Weiler, E.W. & Schröder, J. (1987) *Trends Biochem. Sci.* **12**, 271-275.

Wenck, R.A., Czakó, M. & Márton, L. (1997) *Plant Physiol.* (submitted).

White, F.F., Taylor, B.H., Huffman, G.A., Gordon, M.P. & Nester, E.W. (1985) *J. Bacteriol.* **164**, 33-44.

Wick, M.J., Frank, D.W., Storey, D.G. & Iglewsky, B.H. (1990) *Annu. Rev. Microbiol.* **44**, 335-363.

Widholm, J.M. & Kishinami, I. (1988) *Plant Physiol.* **86**, 266-269.

Xiang, C. & Guerra, D.J. (1993) *Plant Physiol.* **102**, 287-293.

Xu, H., Knox, R.B., Taylor, P.E. & Singh, M.B. (1995) *Proc. Natl. Acad. Sci. USA* **92**, 2106-2110.

Yamazumi, M., Mekada, E., Uchida, T. & Okada, Y. (1978) *Cell* **15**, 245-250.

Zhang, H., Scheirer, D.C., Fowle, W.H. & Goodman, H.M. (1992) *Plant Cell* **4**, 1575-1588.

Ziljstra, C. & Hohn, T. (1992) *Plant Cell* **4**, 1471-1484.

Considerations for Development and Commercialization of Plant Cell Processes and Products

Walter E. Goldstein

Escagenetics Corp., 830 Branston Road, San Carlos, CA 94070-3305, and Walter Goldstein Consulting and Marketing Services, 4 East Court Lane, Foster City, CA 94404-2133, USA

Introduction

There has been substantial progress toward meeting technical and business goals to commercialize products manufactured by plant cell processes. The progress is due to better assessment of commercialization needs, and more knowledge of how to scale-up laboratory procedures and techniques, resolve difficulties, and create opportunities. This progress is examined by reviewing how needs in two different markets might be satisfied by different plant cell culture product possibilities that share a technology in common. The two specific examples are PhytoVanilla™, for the flavor/fragrance markets, and Taxol/taxoids, for the anti-cancer/therapeutic markets. (PhytoVanilla™ is a registered trademark of ESCAgenetics Corporation).

There are several considerations involved when entering a plant cell tissue culture program to develop such products or line of products. Not the least consideration may be a strategic decision to enter a particular market area based on technology in place or technology that one expects to develop; i.e., deciding to enter pharmaceuticals, when your business has been food ingredients, is a significant decision.

The considerations involved in entering such a plant cell tissue culture program are proposed to be those indicated in Table 1 (these considerations are assuredly generic to any new endeavor, especially one that can be fairly described as a pioneering effort). The first consideration involves assessment of market need; that is, will someone want or buy your product if you have it? It is important to ask what your market niche is and to "size-up" your competition. The inability or limitations in making a product by alternative chemical or biochemical synthetic means, trying to insure the consistency and quality of the product, and trying to insure also against social, climatic, and political disasters are all used to justify reasons to develop a plant cell culture process for a particular product; however, in the end, the most important question relates to whether or not someone will buy your product if you have it. You cannot answer this question ahead-of-time, but you have to guess well because you are risking your money. This "educated-guessing" entails a proper balance of a quantitative and qualitative evaluation that is completed in marketing research.

- assessing market needs
- estimating technical/economic feasibility
- developing laboratory procedures leading to stable and productive cell lines
- attaining knowledge of scale-up procedures
  - propagation/product formation
  - recovery/purification

Table 1: Considerations for entering plant cell tissue culture product development

If you believe you will assuredly sell the product if it has the right characteristics and is offered at the right price, then you pass to the stage of assessing feasibility. For technical feasibility, you assess if you can complete development work successfully to make the product in an acceptable time frame. The relevant questions involved are: (1) how much experience do you have in such work; (2) how many programs have you previously conducted in this area; and (3) can you learn from your errors? Having the experience of several plant cell culture development programs, along with all the history of mishaps and mistakes inherently involved in working in such an unknown area, and understanding what may have gone wrong, are critically important to the success of such a program.

In the case of PhytoVanilla™, at inception, this was viewed to be an impossible assignment by many (paraphrasing some comments that may have been made a decade ago, "Vanilla? that is produced in the fruit of the vanilla *orchid* as a consequence of a post-harvest natural *fermentation* process--an impossible assignment in plant cell culture!"). The program turned out to be a difficult task, though PhytoVanilla™ has been shown to be made through plant cell culture, patents have been issued, and there has been progress toward commercialization. The "fringe benefit" that was not fully appreciated at the time is that the teachings of PhytoVanilla™ and other programs along the way have been important to the success of programs conducted later such as taxol (at the beginning for taxol, paraphrasing statements by some, "deriving taxol through plant cell culture of a slow-growing woody plant, a tree, impossible!").

The efforts on each task involved in development of a culture line in a new product area in parallel or subsequent to work on PhytoVanilla™, while not necessarily less challenging or substantially easier than PhytoVanilla™, seemed to be completed with much more assurance and dispatch. Therefore it is important to assess if your experience can be helpful in a plant cell culture endeavor (establishing a "learning curve" has not been easy because too much was and still is unknown in the field to provide guidelines). Therefore, as is the case in such new fields, decisions on entry are based on using best judgment derived from available knowledge, and most-of-all, your "gut" feeling about what lies ahead.

Given that you think somebody will want to buy your product, and that you can complete the technical program, you have to determine ahead of entry that you can make money from the project. This involves an economic assessment of the technology, its translation into commercial production, the costs to clear regulatory hurdles, and the cost involved in developing and entering the market. Questions must be answered and predictions made about yields, recoveries, and processes at the beginning that may be partly conceptual. Basically, your success depends on how good a job has been done in writing the *business plan*.

To answer many of these questions, laboratory procedures for cell culture have to be developed leading to stable and productive cell lines, hopefully, as mentioned, based on what has been previously accomplished. At the beginning of a program, there should be a reasonable expectation that the plant can be cultured or propagated, and that a product can be obtained from the culture in copious amounts based on knowledge as to how to induce or stimulate the cells to produce the product.

Bench scale and analytical procedures have to be developed to assess the presence of the product and the co-products or impurities present. Then, as a separate development task, scale-up procedures have to be developed, involving propagation of the cells and forming the product in copious amounts at a different scale. The procedures are derived from those used in the laboratory, but involve different environmental influences on cell biology which have to be learned and taken into account. Then, at large scale, processes must be developed based on laboratory procedures to economically and consistently recover and purify the product to an adequate and consistent degree. Homework must be completed on several fronts in order to make decisions on entering plant cell tissue culture, and then the work followed up on properly if the decision is made to proceed. PhytoVanilla™ and taxol are appropriate, pertinent examples (and learning experiences) to present aspects involved in plant cell tissue culture development.

PhytoVanilla™ is intended to enter a market comprised of vanilla extract and synthetic vanillin, and be accepted because of quality and price advantages, if such advantages can be realized. Food products such as ice cream, soft drink beverages, and yogurt products are examples of areas where PhytoVanilla™ might find a niche.

Plant cell culture-derived taxoids such as taxol can enter therapeutic markets if they satisfy treatment, product availability, and quality needs. The taxoid products will also be favored if they improve results obtained in treatment of diseases such as cancer or viral infections on their own merits, as a consequence of synergy with other therapeutic agents, and/or because they provide a more favorable low cost alternative for treatment.

The commercial availability of plant cell culture-derived PhytoVanilla™ and taxol will be a consequence of translating laboratory cell biology methods to pilot and manufacturing scale, and meeting economic, consistency, and quality criteria. Problems encountered to be resolved include maintaining asepsis in propagation, adapting cells to production conditions, and adapting methods from analogous culture areas to the special needs of plant cells. Commercialization will result from solving such problems and achieving superior economics as a consequence of process development. Overcoming the problems helps establish a technology that can hopefully lead to other plant-based product opportunities to benefit human health and welfare.

Technology description

Plant cell culture is well-established for many species. Cell lines are initiated by taking sterile cuttings from aseptic plant tissue and exposing tissue to nutrients and plant hormones using serial culture (Staba, 1980; Dixon, 1985; Bu'Lock and Kristiansen, 1987; Shuler and Kargi, 1992; Payne et al, 1991). The plant tissue eventually dedifferentiates and forms what is known as a callus. The callus is propagated until it can grow independently of plant tissue. This process continues until the callus is in a form so that it easily separates into individual cells or aggregates of cells to continue propagation in suspension. The cells to be propagated may be in different states or forms of differentiation. It is more common to propagate them in a state of nearly complete dedifferentiation (callus), though the state desired is one where stability is achieved and the cells are maximally able to produce the product of interest.

The population must be taken to a state dependent on inherent cell biology and environment to which the cells are exposed so that the cells can be propagated consistently. Serial culturing continues in suspension under prescribed conditions to attempt to obtain a consistent and non-segregating culture, suitable for bulk propagation. This can involve investigating a number of conditions for an indefinite time; for example, cell source, cell type, light, temperature, nutrient level, hormone level, and liquid/gas phase composition. Generally, conditions are fixed, and attempts are made to attain a non-segregating cell line, by testing several lines. The desired phenotype that overproduces compounds of interest, or has a reduced doubling time, is obtained by a proper combination of a stable genotype and a complex, but well-controlled environment. Often the combination of parameters is a matter of hit-and-miss, or serendipity.

Conditions can also be varied to investigate many parameters at the same time to develop a new or improved cell line. If many parameters are being investigated, and cells selected entirely by qualitative and empirical observation, the number of parameters is by necessity restricted to those that the investigator can easily manage and still be able to determine which parameter(s) may have resulted in the cell line obtained. Generally, rate and consistency of cell growth are monitored in serial culture to see if the cell line propagates in a stable fashion. If cell selection by empirical observation is augmented by measurement of product formation based on assay, staining procedures, examination of pathway products such as metabolite or enzyme, or DNA and/or chromosomal changes, then more information may be obtained. The methods that can be used are obviously a function of the extent of knowledge at the time, the inherent stability of the cell line, and the techniques available for use.

If cell line stability has been achieved by means of serial transfer, and improvement of a particular characteristic such as production of a particular compound is desired, conditions are imposed to cause the cells to purposely change in a direction to obtain characteristics desired, such as more of a particular product, and the line is then stabilized again as demonstrated by consistency in repeated subculture.

It is apparent that a major difficulty with the technology is that the cell lines will drift, possibly due to less than adequate control of the environment, or due to natural genetic change. Learning how to manage this drift at small scale and maintaining stability as scale-up proceeds is part of what is critical to process development.

The process to develop stable and productive plant cells is part empirical, and has not yet been able to take advantage of developments for other cellular processes where cells are expected to be stabilized by freeze-drying, maintenance on slants, or else in a frozen, cryogenic state. The process could benefit by use of statistical experimental design methods, focused on monitoring some observable characteristic(s) that is (are) reflective of a desirable trait.

Considerations in development of PhytoVanilla™

The vanilla flavor market

The flavor and fragrance business worldwide is approximately US$2 to 3 billion. This is comprised of natural and synthetic products. Use of natural products has been popular; however, such popularity depends on their having better economic value than what is in the market place or having some utility not already provided. The flavor business is approximately US$300 million worldwide, with the vanilla-related flavor business comprising about US$250 million, half synthetic vanillin, and half vanilla extract (Table 2). Vanilla extract is a complex mixture extracted from the vanilla bean, where the major components contributing to the rich characteristic bouquet and flavor are essentially, vanillin and about five others, with some trace components likely important from over 200 components in the bean.

- worldwide market size of US$250 million
- product applications
  - **vanilla extract**
    - ice cream, soft drinks, and extract
    - baked goods and confectionery
  - **synthetic vanillin**
    - ice cream, confectionery, and baked goods
    - soft drinks, flavored milk, and pudding/toppings

Table 2: The vanilla extract/synthetic vanillin market

The many components in vanilla extract have a flavor impact in products which is expressed in terms of flavor units, a somewhat empirical number, where one flavor unit depicts the organoleptic impact of 0.83 lbs. or 13.35 oz. by weight of vanilla beans. Single strength vanilla extract involves extraction of 13.35 oz. of vanilla beans or one flavor unit into one gallon of ethanol. The flavor unit is a qualitative frame of reference to judge flavor impact obtained, for example, by substituting multiple quantities of less-expensive vanillin for more expensive vanilla extract.

In general, substitution by quantity does not substitute for quality; for example, it is generally easy to discriminate between the taste of an ice cream or yogurt formulated with vanilla extract in contrast to one formulated with vanillin. However, the cost of providing vanillin in the form of vanilla extract is considerably more expensive than providing it as synthetic material. Therefore, a useful natural substitute for vanilla extract would mimic vanilla extract quality, but at a much lower price. PhytoVanilla™ is hoped to be such a product, where value would be gained by substituting for vanilla extract where this would be possible based on cost/price and regulations governing such substitution. PhytoVanilla™ would substitute for synthetic vanillin (derived from a by-product of wood processing) when a natural substitute is preferred, and purchasers are willing pay more than they would pay for synthetic vanillin, but still less than they pay for vanilla extract.

Marketing and regulatory aspects complicate the commercialization of PhytoVanilla™. For example, vanilla extract must be used in certain products according to domestic and international regulations such as Category I ice cream (USA Code of Federal Regulations) and in certain flavorful alcoholic beverages in Europe. Second, the growing and harvesting of vanilla beans is a proprietary and valued business in particular locales of the world such as Madagascar, Tahiti, and Indonesia, and the people involved in this business are protective of their interests. Finally, there is a complex international business infrastructure comprising interested parties who are involved in the trading, marketing, distribution, storage, and extraction of vanilla beans.

This network has been built up for some time. To break into this network requires strong justification, influence, and financial means. In many respects, and in many locales, the business is very closely-held. Further, the testing of vanilla extract is very involved, for example to determine components in varied fractions, and to validate that the extract is natural, the real vanilla. Tests using radioisotopes can attempt to ascertain that a source of vanilla or vanillin comes from a natural source derived from by particular biochemical pathway, since certain pathways are expected to be enriched in a particular isotope. The tests involve determining the relative proportion of the radioisotope, carbon 13, to carbon 12, and the relative proportion of deuterium to hydrogen. Because vanillin is so inexpensive in comparison to vanilla extract, parties sometimes attempt to enrich the more expensive natural vanilla extract by adding synthetic, non-naturally-derived vanillin, without necessary communicating that this is taking place (acts of adulteration can be a facet of problems faced by the industry).

Some parties selling vanilla beans from one part of the world may wish to have an advantage against parties selling vanilla beans from another part of the world (this is an interesting subject in itself, since the composition of beans from different parts of the world likely varies in composition of vanillin and other flavor-supporting components). PhytoVanilla™, a natural product derived from plant cells, could be blended into vanilla extract to enrich it and maintain its natural character. This would be of value if the cost of the final product could be lower. Since this product could complicate an already complicated market, which has existed for a long time, it is likely that there are parties who would just as soon not have the product reach the market place, since the product could scavenge business from others. However, this partial replacement of vanilla extract with PhytoVanilla™ would be expected to happen if PhytoVanilla™ met quality needs and if its price offered advantages.

A "win-win" scenario would have a range of PhytoVanilla™ natural flavors increasing the number of flavor and flavor-enhancement possibilities. For example, a vanilla flavor could be tailored for ice cream, another for chocolate bars, another for beverages, and another for yogurt and other health foods. The marketing possibilities are evident if supply of the quality compound can be established at the right price.

Pricing and value relationships for a natural vanilla flavor substitute

Vanilla beans are generally purchased at near US$40 per pound of beans. A particular standard concentration of ethanol extract of the beans, termed single fold extract, about 0.83 lb. of beans per gallon of final extract then is valued at US$34 per gallon. The wholesale-to-retail cost may be typically between US$50 and US$100/gal of extract; therefore, assume a value of US$75 per gal to account for processing and other costs, and markups for bulk extract. At 128 fluid oz. per gallon (a fluid oz. is a standard unit of measure for vanilla extract), a cost of US$75 per gallon of extract becomes US$0.50/fluid oz. of single fold extract (two-fold extract, twice as concentrated, is $1/fluid oz.). If a gallon of ice cream requires about 0.5 oz. of extract, then each gallon of quality ice cream has about $0.25 of extract in it (a substantial cost component of the ice cream which may retail for between US$3.50 and $7.00 per gallon).

Each ounce of single fold vanilla extract has about 100 mg of vanillin flavor components. At $0.50 per fluid oz., then $0.50 per fluid oz. becomes US$0.005/mg of vanillin flavor components, or US$5,000/kg of vanillin flavor components at the wholesale/retail level. At the level of manufacturing cost, a figure of US$2,500/kg is more appropriate. Certain ice cream products use synthetic vanillin which costs about US$7/lb. or US $ 15/kg. To attempt to obtain the flavor impact of ice cream containing the bean extract, more synthetic vanillin is used relative to the amount of extract in the recipe, e.g., three to six times or more synthetic vanillin.

Therefore, if a gallon of ice cream requires 0.5 oz. of bean extract or 50 mg of vanillin flavor components, then three to six times this figure results in use of 150 to 300 mg of vanillin. The relative cost of vanilla extract to synthetic vanillin use in this ice cream example application is then US$2,500/kg for vanilla extract divided by $15/kg (the cost of synthetic vanillin), divided again by the factor of 3 to 6 weight of vanillin substituted per weight of vanilla extract (more synthetic vanillin is formulated into an ice cream product to try and obtain the flavor impact of vanilla extract). The result of the calculation is the effective cost to use synthetic vanillin instead of vanilla extract, or US$50 to 100/kg synthetic vanillin (the cost ratio of vanilla extract to synthetic vanillin is 25 to 50). Obviously, if synthetic vanillin can provide adequate flavor impact in the product, the cost advantage for its use is substantial.

However, if the synthetic vanillin does not give the required quality unless partially blended with vanilla extract, then the effective cost of the vanillin/vanilla extract flavorant increases for the particular application. The blending required depends on the results of taste tests involving actual product (e.g., ice cream or beverage) formulations. For any particular application, the resulting blend ratio and associated cost for the blend for a range of applications or a particular application becomes the relevant reference figure or "straw man" for substitution or replacement of a portion of the extract with an extract containing PhytoVanilla™ from plant cell culture. Replacement in any particular application requires assessment of flavor impact required, and the cost for the flavorant that the application will tolerate, among other requirements. It is clear that the arena of vanilla economics has its own special brand of complexity.

Potential development and economic value of PhytoVanilla™

It is useful to assess a potential market for PhytoVanilla™ and provide an example of a plausible and also positive event. The total vanilla market (extract and synthetic vanillin) is about US$250 million. PhytoVanilla™ can enter this market by replacing vanilla extract or synthetic vanillin in existing product applications or else by expanding the market through new product growth. Assume that there is the potential to capture 20 % of the vanilla extract and 5 % of the synthetic vanillin markets, and that market growth could be responsible for part of the sales instead of the product simply cutting into existing markets. Assume, also, a 1:1 replacement of vanilla extract at lower cost and replacement (arbitrarily) of ten parts of synthetic vanillin with one part of PhytoVanilla™ (the actual replacement level has to be determined on a product-by-product basis).

Then, US$25 million of annual vanilla extract flavor sales at US$5,000/kg flavor solids, or 5,000 kg is replaced; and US$6 million of synthetic vanillin at US$15/kg is replaced (at the 10 parts synthetic vanillin to 1 part PhytoVanilla™ level, 30,000 kg per year of synthetic vanillin is replaced). As a consequence, in this example, 45,000 kg per year PhytoVanilla™ business would result. The pricing of this product would depend on the application, and for the producer/seller, is set based on achieving certain costs through technology advance; however, it is apparent that if such an example becomes reality, then a substantial multi-million dollar business results. Such a business is worth entering if the return on the investment to enter the business is justified.

The economics of plant cell culture processes were estimated by the author and others in the past (Goldstein et al, 1980; Sahai and Knuth, 1985). The raw material, labor, utility, and capital-related charges (for example, number of fermentors and processing equipment needed) are highly yield dependent. Multiples of large high quality aseptic fermentors would be required, ca. 40,000 liters each, and substantial capital cost for new equipment, lease or tolling costs would be required to meet a production level of 45,000 kg per year.

• consisting of over 200 **known** components

principal components contributing to the flavor/organoleptic properties are:

- vanillin
- p-hydroxybenzaldehyde
- vanillic acid
- p-hydroxybenzoic acid
- acetic acid
- p-hydroxybenzyl methyl ether
- benzyl alcohol
- traces of other compounds

Table 3: Vanilla extract composition, an ethanol extract of vanilla beans

In actuality, the process, still in the midst of piloting and scale-up, would be scaled up in only a few such large fermentors initially to fully develop the technology at large scale before retrofitting of many such fermentors is completed. The yields needed are within the realm of possibility to be obtained in order to meet cost targets so that plant cell culture manufacturing produces PhytoVanilla™ at a substantially lower cost than the cost of vanilla flavor extracted from beans. It is necessary, of course, that the quality of the flavor produced matches that achieved for vanilla extract (Table 4). The product would be even more desirable if flavor nuances are discovered to allow identification of new market applications for the PhytoVanilla™ flavor system. The opportunity lies in determining the empirical and practical correlation by application between the product composition, e.g., Table 3, and organoleptic attributes (Table 4).

- woody, rummy, balsamic, sweet
- buttery, candy, creamy, floral
- phenolic, cooked, vanillic

Table 4: Descriptive organoleptic attributes of vanilla extract

Whether such new flavor systems will result to spur market growth remains to be seen. The composition of the current PhytoVanilla™, although not compositionally identical always to vanilla extract (Table 3), exhibits the same type of flavor qualities, and contains components typically found in foods and food ingredients.

The technology achievement required for PhytoVanilla™ in production involves matching productivity and maintaining asepsis in large scale systems comparable to that achieved with systems at small scale; this involves propagation of cells that grow relatively slowly (doubling times involving several days). Such doubling times contrast significantly with cells that one is normally more used to when dealing with fermentation systems (mammalian or fungi cells involve doubling times less than 24 hours, and bacterial cells double in less than one hour). Such scale-up challenges are also being undertaken on another system, taxol, a program that began based on the technology developed for PhytoVanilla™, and this subject is reviewed later in this paper. A summary of the technology achievements required for PhytoVanilla™ are summarized in Table 5.

Certainly, the same type of manipulation of culture used to change the type of molecules (taxoids) appearing in a taxol broth, is also possible for a broth containing PhytoVanilla™, and this could be a path to new flavor systems of added commercial value. The potential exists for a multi-million dollar profitable product, with additional work and investment being necessary to bring this about.

Developing the technology and being able to apply it more easily the next time for other applications could establish new business opportunities if the problems inherent in the technology to advance it can be solved. Many of the advances in plant cell culture in general are a consequence of the work on PhytoVanilla™, and more advances based on PhytoVanilla™ are occurring on taxol

---

• demonstrate yield necessary at laboratory and pilot plant scale to satisfy economics required when the process is in production

• demonstrate product quality needed in different applications such as the example discussed

• demonstrate the scaleability of cell systems in production fermentors (ca. 40,000 liters), involving:
a) maintenance of asepsis.
b) stable propagation of the cells.
c) product yield and quality after recovery and purification, matching results at smaller scale.

Table 5: Technology achievements required for PhytoVanilla™

Considerations for taxol derived from plant cell culture

Taxol costs

Taxol derived from bark is estimated to cost US$500,000/kg and this cost is reported to be dropping. Deriving a taxol precursor from a needle extract and conversion of the precursor to taxol is expected to result in a lower cost taxol. The cell culture process has the potential to cost the least of all, perhaps less than half of the US$500,000/kg figure depending on the production volume. The present cost of taxol to treat a patient is approximately US$10,000 per patient per year and this might involve 1 to 2 grams. Therefore at the hospital pharmacy level, the product is worth US$5,000,000/kg. The potential need for taxol ranges from several hundred kilograms up to quantities exceeding 1,000 kg per year. Therefore, the business value of this product area is considerable.

Sourcing, mode of action, and product development

The taxol program at ESCAgenetics Corporation emerged as an identified opportunity based on the progress that had been made on PhytoVanilla™. There was a shortage of the compound and the source (yew trees) of taxol was not acceptable as a permanent solution for many reasons, such as ecological concerns and cost. The technology developed for PhytoVanilla™ was adapted to taxol, and cell lines based on the Taxus genus were developed. A breakthrough, confirming *de novo* synthesis of taxol occurred after work started. This anti-cancer drug, currently approved for ovarian and breast cancer is supplied from extracts of the bark of yew trees, and the next source discussed is use of a taxol precursor extracted from yew needles and subsequent conversion of this precursor to taxol using a patented process.

Plant cell culture derived taxol, which began as a low probability source of supply, is emerging as much more of a possibility if cost targets can be met. Currently, the yield of taxol in culture well exceeds by many multiples taxol content in bark or needles to be extracted or converted.

Taxol acts by promoting the assembly and inhibiting the disassembly of microtubules in cells during mitosis, and in effect, slows the growth of cancer cells in comparison to normal cells, possibly inviting their destruction by the immune system. Taxol is a complex molecule, molecular weight of 853, with 11 chiral centers and very difficult to synthesize. Even though the remarkable feat of its synthesis has been recently achieved, chemical synthesis is unlikely to be commercially practical.

In order to commercially introduce a plant cell-derived taxol, the process to produce taxol by plant cell culture must be fully developed, pass cGMP criteria, and the product proved to be analytically-equivalent and bioequivalent to the product already approved. Therefore, the product must be purified to an acceptable degree using a validated process. Plant cells containing taxol are easier to process than leaf or bark, so it is expected that obtaining a sufficiently purified taxol should be possible, one the plant cell propagation and taxol formation process has been fully developed for use at commercial scale. The product will be marketed as a generic drug where the product has already been approved. Such marketing, even with a product showing cost advantages, is restricted by statutes in certain countries such as the USA that delay the introduction of generic products.

Pharmaceuticals from plant cell culture

Developing the technology to produce taxol and the manipulations involved in culture led to the discovery of other taxoids and potential new drugs in the broth. The establishment of this technology for production of other plant products, or as a basis to screen for new plant products, promises to open new vistas in medicine. Such opportunities can result by taking advantage of ethnobotanical, plant-based medicine practiced in many cultures, combined with medicine practiced in developed nations. For example, there may be similarities and analogies between mixtures produced by knowledgeable shamans, and delivery systems developed by pharmaceutical scientists that can lead to advances in drug effectiveness and reduced side-effects. Further, molecular analogs discovered in plant-based sources when compared to compounds discovered through synthetic chemistry and biosynthetic means may result in new directions for medicine. Therefore, the technology which was developed while trying to create a new way to make a food flavor ingredient is being used to help make drugs that may help cure disease and make those more comfortable who have such diseases.

Technology and needs for plant cell-derived products

Development of the PhytoVanilla™ process and the taxol process faced similar obstacles. Plant cells grow slowly, doubling every few days for PhytoVanilla™ to about a week or so for Taxus. Such slow growth invites microbial infestation unless asepsis practices are well developed, people properly-trained, and appropriate scaleable equipment specified. Further, the amount of dilution possible for plant cell culture is far removed from what is possible for microbial or mammalian systems. Plant cells often cannot be diluted more than 1:6 or 1:4, and this means many multiple stages of propagation are required, inviting more chances for error or contamination. Cell line dilution must be maximized as far as possible without making the system unstable; otherwise, production trains involving increasing large fermentors employing steps such as variable volume techniques must be used to reduce the number of fermentors in the train to lessen the risk of failure. This is valuable technology, but requires great attention to detail and care in process control and monitoring to be successful.

Plant cells appear to have the tendency to manipulate the environment they are in if conditions are not what they should be. If the conditions are far removed, or the cells are "shocked", growth may, for example, slow or stop. Some of the conditions relate to proper control of gas or liquid phase conditions. The cells need an adequate oxygen supply, but this cannot be provided in a manner that either imposes excessive mechanical shear or else bubble related shear (in effect "shot-peening" the cells). Therefore, agitator and gas distribution design is critical as is control of broth partial pressures. Such control is essential at all stages of scale-up, from laboratory to pilot, and to large scale.

The control of conditions will vary depending on the scale of operation (for example, increased liquid head changes the partial pressure of dissolved components that can influence the culture). Aspects of the control required are predictable from related areas such as microbial and mammalian fermentation; however, the way to accomplish control is often not predictable at all. It is clear that cell biology must be developed from the beginning to understand which parameters are important and to obtain a cell line that will propagate consistently.

Part of the problem can be the heterogeneity of cell line and the propensity of the cells to drift from their particular state of non-differentiation. Such drift may be partly manageable through process control and knowing the target value and allowable range of parameters such as nutrients, plant hormones, and gasses and how to maintain them within a desired range during the course of the run.

This is a complex area, requiring an ever-increasing body of empirical data, good analytical methods, and the availability of computer-based monitoring and sensors. The field could benefit from a better understanding of the underlying theory of plant cell tissue culture, and application of statistical experimental design and process control methods to augment visual observations of experts responsible for selecting and developing cell lines.

This field also requires people with a lot of patience, determination, follow-through, and vision. In that sense, it does not differ from other fields where pioneering is required. Plant cell tissue culture, PhytoVanilla™ and taxol have been the basis for ESCAgenetics' entry to pharmaceutical discovery, and manipulation of plant cells in culture, and plant sources in general, provide new promise as a means to discover new drugs and take advantage of plant-based medicine. The program has evolved into one that is attractive and worthy from both humanistic and financial standpoints.

References

Bu'Lock, J. & Kristiansen, J. (1987) In: *Basic Biotechnology.* Academic Press, London, UK.

Dixon, R.A. (1985) *Plant Cell Culture, A Practical Approach,* IRL Press, Oxford.

Goldstein, W.E., Lasure, L.L. & Ingle, M.B. (1980) In: *Plant Cell Culture as a Source of Biochemicals,* Staba, E.J. (ed.), CRC Press, Boca Raton, FL.

Payne, M.L., Bringi, V., Prince, C.L. & Shuler, M.L. (1991) *Plant Cell Tissue Culture in Liquid Systems.* John Wiley & Sons, New York.

Sahai, O. & Knuth, M. (1985) *Commercializing Plant Tissue Culture Processes: Economics, Problems and Prospects. Biotechnology Progress* (Vol. 1, No. 1).

Shuler, M.L. & Kargi, F. (1992) In: *Bioprocess Engineering: Basic Concepts.* Prentice Hall publ. Englewoodd Cliffs, NJ.
Staba, E.J. (1980) *Plant Cell Culture as a Source of Biochemicals,* CRC Press, Boca Raton, FL.

DNA Diagnostics in Horticulture

Wm. Vance Baird[1], Albert G. Abbott[2], Robert Ballard[2],
Bryon Sosinski[2] and Sriyani Rajapakse[2]

[1]*Department of Horticulture, Box 340375 and* [2]*Department of Biological Sciences*
Clemson University, Clemson, SC 29634, USA

Introduction

The utilization of DNA-based genetic markers has signaled a new era in genome analysis. The genetic variation detected by these markers can be used to identify loci affecting qualitative or quantitative traits, to monitor loci during breeding and selection programs, and to access taxonomic relationships. They are particularly well suited to chromosome mapping in species where crosses are difficult to make. This is because DNA polymorphisms are more abundant than conventional phenotypic and biochemical markers, allowing a molecular linkage map to be developed in a single segregating population.

Molecular markers offer many other advantages over conventional phenotypic markers in that they: 1) are developmentally stable; 2) are detectable in all tissues; 3) are unaffected by environmental conditions; 4) generally lack epistatic and pleiotropic effects; and 5) provide a choice of codominant or dominant marker subclasses (Botstein et al, 1980; Helentjaris et al, 1987). The theoretically unlimited supply of molecular markers has the advantage of increasing the probability of tagging quantitative, multigene traits, as well as those controlled by single genes. This potential marker abundance is important for developing molecular genetic linkage maps where a dense, uniform distribution is desired. Molecular genetic maps allow the development and use of indirect selection schemes known as marker-assisted selection (MAS) for germplasm improvement. Such strategies strive to increase precision and efficiency in the manipulation of both qualitative and quantitative traits.

Construction of molecular genetic maps have allowed the isolation of genes involved in pathogen resistance or those controlling growth habit and response to environmental stresses (Mindrinos et al, 1994; Bent et al, 1994; Martin et al,

1993; Arondel et al, 1992). Other applications include: 1) evaluating germplasm collections for genetic variability; 2) determining the level of heterozygosity and homozygosity in breeding stock; 3) monitoring the introgression of important genes from wild species; and 4) reducing the effect of linkage drag (Young and Tanksley, 1989; Tanksley et al, 1989).

The development of molecular genetic maps offers a number of advantages to plant breeders. For example, screening progeny for the presence of desirable alleles (e.g., pest resistance) often requires field evaluation. Field testing limits the size of the breeding program because of the demand it creates for space, time, and personnel. In the case of plants failing to reach maturity in one growing season, field evaluation becomes even more crucial in terms of effort applied and resources allocated to a breeding program. When linkages between DNA marker loci and beneficial traits are sufficiently tight, it is possible to predict, soon after germination, whether beneficial genes will be present in the progeny. This early assessment of genetic potential and subsequent culling will reduce production costs and likely expedite new cultivar development and release. Also, when codominant genetic markers are used, one can detect the presence of recessive alleles (i.e., identify heterozygotes), which may be important in the early stages of many breeding programs.

This overview of the use of DNA diagnostic methods in horticultural science will cover crops such as fruit trees, small fruits, vegetables, and ornamentals. By design, grain and most legume crops have been omitted (see previous volume in this series), as well as the use of DNA diagnostics in turf (see next chapter). The studies reported here use molecular biological methods and genetic markers to address questions of variety improvement, resistance to biotic and abiotic stresses, phylogenetic relationships of cultivated crops to their wild relatives, and cultivar identification. A number of areas where DNA analytical methods are used in horticulture, but are not strictly related to DNA diagnostics and their use in crop improvement will not be discussed here.

DNA markers

Germplasm evaluation and enhancement efforts have always needed an efficient, reliable, and direct method for detecting, characterizing, and following the inheritance of genetic variation. Ideally, such a method would be both rapid and objective in its application and utilize stable, quantifiable marker characters. These requirements are currently best satisfied by DNA markers.

The first broadly applicable DNA-based markers detected variation in the length of DNA sequences revealed by digesting genomic DNA with a restriction endonuclease and screening the resulting fragments by Southern hybridization with a DNA clone (Kan and Dozy, 1978; Wyman and White, 1980; Botstein et al, 1980). Therefore, these genetic markers are referred to as restriction fragment length polymorphisms (RFLPs). These molecular probes are either random genomic clones, cDNAs, or synthetic oligonucleotides. This analysis system, in which the markers are codominant, has transformed the genetic characterization of plant germplasm, and is contributing significantly to map-based selection in many breeding programs. Despite this utility, RFLP analysis has some limitations: 1) the process is time-consuming and labor intensive; 2) marker management and dissemination is tedious; 3) relatively large amounts of DNA are required; and 4) automation of the procedures is difficult. Also, for many highly inbred crops, RFLPs reveal a relatively limited number of alleles per locus.

Although a number of other methods have been devised to detect DNA polymorphisms due to single base changes, these methods tend to be even more labor intensive than the standard RFLP approach, and are currently used only in somewhat specialized circumstances. These methods include two-dimensional gel electrophoresis and denaturing gradient gel electrophoresis.

A major advancement in germplasm analysis and mapping occurred with the development of the polymerase chain reaction (PCR; Mullis et al, 1986). PCR-based methods allow the rapid detection of DNA polymorphisms from many individuals or pooled samples. Several important PCR-based methods have been developed, and used extensively in the analysis of plant genomes. In general, these methods employ a series of single short (e.g., ≤10 nucleotide) oligonucleotide primers of arbitrary sequence to amplify one or more regions of the genome, and produce a DNA banding pattern unique to each primer. This pattern can be visualized directly, without the intervening need for blotting and labeled probes, in ethidium bromide stained agarose gels. This method, now usually referred to as random amplified polymorphic DNA (RAPD), was developed independently in two laboratories (Welsh and McClelland, 1990; Williams et al, 1990). This approach has the advantage that no prior knowledge of the organism's genome (i.e., sequence data) is needed to initiate the analysis. RAPD analysis has quickly become an important tool in plant genetics and breeding.

Initially, PCR-based markers suffered from non-reproducibility. At least part of the inconsistency was attributable to the fact that these methods did not require or even use knowledge of any specific region in the genome to design the amplification primers. By cloning and sequencing RAPD products of particular importance, specific amplification primers could be designed that would consistently amplify only the region(s) of interest. These sequence characterized amplified regions (SCARs) use a pair of unique primers (approximately 20 nucleotides) coupled with restriction digestion to visualize genomic regions of interest and screen for polymorphisms, thus improving the reproducibility of the PCR reaction and converting dominant markers to codominant markers (Paran and Michelmore, 1993).

A second PCR-based method, DNA amplification fingerprinting (DAF), involves the use of shorter primers, reducing the stringency of primer-template annealing and visualizing the amplification products on silver stained polyacrylamide gels (Caetano-Anollés et al, 1991; Bassam et al, 1991). DAF produces detailed and relatively complex DNA profiles suitable for genetic analysis in a wide variety of organisms including plants. This method already has a number of refinements, and has found an important role in cultivar identification.

A third method utilizes very short simple sequence repeats (SSRs) of extremely high polymorphic content, the so-called "microsatellites" (Litt and Luty, 1989; Weber and May, 1989). The study of these sequences will likely revolutionize the molecular analysis of plant genomes as it is doing for other eukaryotic organisms (Dietrich et al, 1992; Weissenbach et al, 1992). Already, a number of reports have appeared demonstrating the utility of microsatellites in plant systems (Condit and Hubble, 1991; Akkaya et al, 1992; Zhao and Kochert, 1993; Morgante and Olivieri, 1993). These dispersed tandem repeats have a basic unit of less than or equal to 6 basepairs. Once microsatellite loci have been identified, cloned and sequenced, a pair of unique primers flanking each repeat is synthesized and used to identify the codominant, simple sequence length polymorphisms (SSLPs; Tautz, 1989). This is usually accomplished by size fractionating the labeled amplification products on denaturing polyacrylamide gels. This class of markers is analyzed as sequence tagged sites (STSs; Olson et al, 1989) for microsatellites; or STMS (Beckmann and Soller, 1991).

A recently developed PCR-based method is termed "amplified fragment length polymorphism" analysis, or AFLP (Vos et al, 1995). The template for this method is double restriction nuclease (RE) digested genomic DNA to which RE-specific adaptors are ligated. Then a series of unique RE-specific primers with one to three additional 3'-nucleotides are used to selectively amplify a relatively large number of products (e.g., about 50). These amplification products are separated in denaturing polyacylamide gels, and can be visualized by autoradiography, automatic fluorochrome detection or silver staining. This proprietary technology holds great potential for molecular linkage map development and DNA "fingerprinting" applications. For example, in potato AFLPs have been used to construct a high resolution map of an interval of chromosome V, flanked by two RFLP markers, that contains the genetic locus (*R1*) for resistance to *Phytophthora infestans* (Meksem et al, 1995).

Linkage analysis and map construction

The effectiveness of a DNA marker for use in MAS or map based gene cloning depends upon the marker's proximity (or linkage), in terms of recombination rate, to the locus of interest. The stronger the linkage or co-inheritance between a gene and a marker, the more effective selection based on that marker will be. Linkage is determined from the percentage of recombination between the gene and the marker in a segregating progeny. Generally speaking, recombination frequency (r) of <0.25 (indicating that in less than 25 % of the individuals recombination between the gene and marker has taken place) is considered sufficient to show linkage. Recombination frequency is also the basis for calculating the genetic distance between the gene and the marker. If r = 0.25, then the genetic distance between them is said to be approximately 25 centiMorgans (cM). For use in MAS, a marker should be within a few cM of the gene.

Linkage data for pairs of markers obtained from a mapping population are used to identify specific linkage groups, which correspond to the chromosomes. Order of the markers in a linkage group is calculated by the multipoint analysis using maximum likelihood or simulated annealing approaches. Several computer software programs are available to calculate linkage between gene loci and markers in segregating mapping populations, and order markers in linkage groups. Some of the more widely used software packages include: 1) MAPMAKER and MAPMAKER/QTL (Lincoln et al, 1992a,b); 2) JOINMAP (Stam, 1993); 3) GMENDEL (Liu and Knapp, 1990); 4) MAPRF (Ritter et al, 1990); 5) LINKAGE-1 (Suiter et al, 1983); and 6) RI Plant Manager developed by Kenneth Manley at the Roswel Park Cancer Institute (Buffalo, New York, USA). These programs vary in the approaches used in generating maps, kinds of segregating populations that can be analyzed, as well as the computer operating systems required for use.

Horticultural crops

Fruit trees

The introduction of molecular marker systems into the study and application of genetics has improved fruit tree breeding. Genetic mapping in these species has traditionally been relatively difficult because of long generation time. Due to the abundance and ease of identification of molecular markers, many fruit tree breeding programs have readily developed marker databases for genetic map construction, fingerprinting, and phylogenetic studies. The best genetically characterized fruit tree species are in the genera *Prunus* and *Malus*. For the more economically important species such as peach (*Prunus persica*) and apple (*Malus domestica*), a number of research groups worldwide are making significant advances in developing molecular marker databases. In the following section, we present the current status of molecular diagnostics in key species of each genus. In addition, we summarize initial efforts for the development of molecular marker databases in other important fruit tree species.

Prunus

Peach (*Prunus persica*) is the best genetically characterized species in the genus *Prunus*, and one of the best genetically characterized fruit crops overall (Mowrey et al, 1990). Peach is a diploid species, 2n = 16, (Jelenkovic and Harrington, 1972) with a remarkably small genome containing approximately 0.30 ± 0.02 pg of DNA per haploid nucleus, or about 3×10^8 bp (Baird et al, 1994).

Structurally, little is known about the genome of peach and its organization. For example, no cytogenetic markers have been identified for the somatic chromosomes of peach, although approximately 25 morphological (Monet et al,

1985) and 10 isoenzyme traits (Monet, 1989; Mowrey et al, 1990) have been described. Unfortunately the latter group has shown consistently little polymorphism in this primarily autogamous species (Arulsekar et al, 1986a,b; Durham et al, 1987; Mowrey et al, 1990).

Surprisingly, among the more than 30 simply inherited traits, only three pairs are known to show linkage (Hesse, 1975; Monet et al, 1985; S. Mehlenbacher, pers. comm., Chaparro et al, 1994). Relatively few examples exist where more than just a few monogenic traits have been incorporated into individual lines to facilitate genetic studies. Also, the use of tissue culture and somatic cell genetics (Roth and Lark, 1984) to study peach would be very limited because few of the available genetic markers display a detectable phenotype in tissue culture. These facts combined with a long generation time, preclude development of a genetic linkage map for commercial peach using conventional methods. Recent reports, however, have clearly demonstrated the utility of molecular markers to make rapid progress in map development and gene tagging in peach (Abbott et al, 1990, 1991, 1992; Eldredge et al, 1992; Chaparro et al, 1994; Rajapakse et al, 1995).

We have constructed a map consisting of 47 linked markers, including RFLP, RAPD, and morphological traits, which identify the eight linkage groups and cover 332 cM of the peach genome with an average spacing of 8 cM (Rajapakse et al, 1995). This map is developed from segregation analysis of molecular markers in 71 F_2 individuals derived from the self-fertilization of four F_1 individuals of a cross between 'New Jersey Pillar' and KV 77119. This cross segregates for genes controlling canopy shape, fruit flesh color, and flower petal color, size, and number. Chaparro et al (1994) reported a map consisting of 83 RAPD, one isoenzyme, and four morphological markers for peach covering 15 linkage groups. They report RAPD markers flanking the red-leaf and malate dehydrogenase gene loci.

For a map to be useful in tree breeding, where segregating populations are not easily generated and agronomically important traits are not expressed in the juvenile condition, molecular markers should be easily generated and/or transferred to crosses that are currently available. From our peach data, we have found that RFLP markers are ideal as "anchor loci" between crosses. We have examined three widely different peach pedigrees and have shown that 40 to 55 % of the markers in one cross detect polymorphic loci in the other crosses. This degree of shared polymorphism has not been observed for RAPD loci where we find approximately only 10 % of RAPD markers are transferable to other crosses (Ballard, unpublished results). In addition, we have utilized our markers in DNA fingerprint analysis of 34 cultivars of peach and have demonstrated that only 9 markers are required to distinguish these cultivars (Rajapakse et al, 1995).

Similarly, researchers at the Istituto Sperimentale per la Frutticoltura, Rome, Italy, have utilized RAPD markers to identify peach and nectarine cultivars. In this study, six primers have been used to identify 29 peach and nectarine genotypes (Quarta et al, 1994). Currently there is a European research consortium, headed by Dr. P. Arus, consisting of laboratories in Spain, France, Italy, and England with the goal of obtaining a highly saturated genetic map for peach. Cooperative efforts of this magnitude will certainly speed the use of molecular diagnostics in this important species.

In addition, heterologous gene probes and random RFLPs are being used to investigate the genetic components of mesocarp development and tree growth habit in peach by employing interspecific *Prunus* hybrids (e.g., almond x peach) (Foolad et al, 1995). Ultimately, a low resolution consensus map appropriate for use with other *Prunus* species will be constructed.

Molecular markers in peach are also useful for characterizing plant material propagated or regenerated from tissue culture. Hashmi et al (1992) identified somaclonal variation for resistance to root-knot nematode *Meloidogyne incognita* in 'Redhaven' and 'Sunhigh' regenerated from culture. These regenerants were screened for RAPD markers and several primers revealed polymorphisms between regenerants and their tissue culture explant-cultivar (Hashmi, 1993). This research supports the genetic basis of somaclonal variation. The need and use of DNA markers will increase as genetic transformation becomes an important component of stone fruit improvement programs (Ye et al, 1994: Scorza et al, 1994).

Molecular map construction in almond (*Prunus dulcis/P. amygdalus*) is underway at the Plant Genetics Department IRTA, Barcelona, Spain, utilizing crosses between almond varieties and interspecific peach x almond progenies. Currently, 85 loci consisting of 78 RFLPs and seven isoenzymes have been identified in an F_1 segregation analysis and have yielded two maps: one containing 66 loci assigned to eight chromosome groups, and the other map containing 44 loci on seven groups. Twenty-five markers are common to both maps (Messeguer et al, 1994).

Malus
In apple, (*Malus domestica/M. pumila*) Matsumoto et al (1993) used ribosomal genes to identify and separate *Malus* species into seven groups based on unique rDNA patterns. Mitochondrial DNA polymorphism, revealed by RFLP analysis, has also been used to place 71 apple cultivars and landraces into four distinct groups that correlate well with geographic origins of apple (Ishikawa et al, 1993). Harada et al (1993) demonstrated that apple species could be fingerprinted by RAPD analysis in combination with restriction digestion of the amplified products. In addition, these investigators demonstrated that five primers could be used to assign paternity to one of six putative candidates for an apple cultivar having an unknown pollen parent.

For *Malus*, several groups are constructing molecular genetic maps. At the Horticulture and Food Research Institute of New Zealand, researchers have developed a map based on 160 molecular markers comprised of RFLPs and RAPDs. In addition, they have assigned by *in situ* hybridization, key RFLP loci to particular chromosomes (Gardiner et al, 1994).

Utilizing F_1 progeny of a cross between the two cultivars, 'White Angel' and 'Rome Beauty', a molecular linkage map has been reported by Hemmat et al (1994). The 'White Angel' map consists of 253 markers arranged in 24 linkage groups extending over 950 cM. The 'Rome Beauty' map consists of 156 markers on 21 linkage groups. Using the 'White Angel' map as the standard, linkage groups homologous to 13 'White Angel' linkage groups were identified in 'Rome Beauty'. Their consensus map for apple has 360 markers spaced at 10 to 15 cM. A vast majority of these markers are RAPD markers, which may not be transferable to other crosses. These researchers have determined that only 10 - 20 % of these amplified fragments were common to both parents. However, this may be an underestimation of the common polymorphisms for other apple cultivars since many crosses have a common parent (Hemmat et al, 1994).

RAPD markers were employed in a phylogenetic analysis, using parsimony, of 25 apple rootstocks (Landry et al, 1994). The derived cladogram was in perfect agreement with the true genetic relationships of the commercial varieties and their F_1 progenies. In addition, a DNA fingerprint system, which allowed the identification of the 25 rootstocks, was developed based on 13 RAPD loci amplified by five RAPD primers.

Citrus

In *Citrus*, molecular diagnostic studies have been applied to cultivar and species identification and to molecular genetic map construction. Matsuyama et al (1993) demonstrated that M13 minisatellite sequences could be used to identify citrus cultivars through PCR and restriction digestion analysis. In addition, they reported variation in *Citrus* cultivar rDNA organization detectable by RFLP analysis.

Yamamoto et al (1993) have reported on the investigation of cytoplasmic genome organization through RFLP analysis of mitochondrial and chloroplast genomes in *Citrus*, *Poncirus*, and *Fortunella*. In this phylogenetic study, *Citrus* was divided into three major groups: pummelo, mandarin, and citron, and the pummelo group was further divided into two subgroups: pummelo and yuzu. *Poncirus* was clearly distinguished from *Citrus*, however, *Fortunella* was not, which is consistent with previously published phylogenetic studies using numerical taxonomic, biochemical, and molecular systems.

The use of PCR-based molecular diagnostics has been reported by Omura et al (1993), where they demonstrated that single and double primed PCR reactions could be used to identify species among the mandarins. In addition, these markers segregate in a Mendelian fashion and could be used in map construction.

Map construction in *Citrus* has been reported for an intergeneric cross between *Citrus* and *Poncirus*. The core map contains 53 codominant isozyme and RFLP markers, and 106 RAPD markers covering 530 cM on nine linkage groups. Using bulk segregant analysis, RAPD markers diagnostic for citrus tristeza virus resistance were identified. This map has been extended to cover 1,192 cM (70 to 80% of the genome) and will eventually be used to identify quantitative trail loci (QTLs) for cold- and salt-tolerance. Three loci detected by a cold-responsive cDNA have been mapped (Cai et al, 1994).

Theobroma

In cocoa, *Theobroma cacao*, molecular diagnostic approaches utilizing RFLP analysis with rDNA, mitochondrial, chloroplast, and cDNA sequences have been used to examine the phylogenetic relationships of the three major morpho-geographical groups designated Criollo, Forastero, and Trinitario. Results of these studies support the presence of two distinct subspecies, *T. cacao* ssp. *cacao* and *T. cacao* ssp. *sphaerocarpum*. Forastero is split into Upper Amazon and Lower Amazon Forastero, while Criollo has differentiated independently on the other side of the Andean Barrier. Trinitario is a result of hybridization between Forastero and Criollo (Laurent et al, 1993a,b; 1994).

In addition, a genetic linkage map of *T. cacao* has been constructed utilizing a combination of isozyme, RFLP, and RAPD loci. The map contains 193 molecular and isozyme markers covering a total length of 759 cM with an average spacing between markers of 3.9 cM.

Oleo

Despite being one of the more economically important tree species, olives (*Oleo europaea*) have only recently been characterized at the molecular level. Previous efforts at molecular diagnostics have focused on allozyme markers in pollen and leaves where it has been demonstrated that cultivars of olive are easily distinguished on this basis (Pontikis et al, 1980; Ouazzani et al, 1993). Efforts to construct a molecular marker database to identify, isolate, and manipulate genes important to oil production, and to develop genetic transformation systems have been recently initiated at the CNR-Istituto Di Recerche Sulla Olivicoltura, Perugia, Italy (L. Baldoni and M. Mencuccini, pers. comm.).

Mangifera

Molecular markers have also been applied to the problems of phylogenetic relationships between mango (*Mangifera indica*) and other *Mangifera* species. Schnell and Knight (1993) have utilized RAPD analysis to define the closest species to *M. indica*. Utilizing 10 primers, 109 bands were resolved for 9 species of *Magnifera*. Their data supported some but not all of the traditional groupings based on anatomical and morphological characters.

Carica

The control of the most destructive disease of papaya (*Carica papaya*), papaya ringspot virus (PRV), has prompted the use of RAPD markers to construct a genetic linkage map for papaya (Sondur et al, 1996). This low resolution map (~10 cM average spacing) has 96 markers and identified linkage of DNA markers to a number of quantitative traits of horticultural interest (e.g., PRV titer, PRV symptom severity, flowering time, fruit weight, and sterility).

Musa

In banana (*Musa* sp.), plants having undesirable phenotypes, e.g., dwarfism, can result from somaclonal variation after transplanting meristem cultures. Eighty "normal" and dwarf *in vitro* plants were compared, using RAPD markers, for early detection of dwarfism (Damasco et al, 1994). These researchers identified a single primer that consistently amplified a 1.5 kb band in "normal" plants, which was absent from all dwarf plants tested.

Small Fruits

Despite their regional or national importance in the USA and world economies, only a few small fruit crops have been examined using methods of DNA analysis and characterization. Many of these studies are broadly focused on cultivar identification and/or determining phylogenetic relationships, while others seek to build a molecular genetic map for use in ongoing breeding programs.

Vitis

Grapes, *Vitis vinifera*, important for wine, raisins, and as a table fruit, are under study in the USA, France, Australia, and Japan. Bourquin et al (1993) used RFLPs in taxonomic studies of eco-geographic groups of *V. vinifera*. Other recent investigations have successfully used sequence-tagged microsatellites for mapping and fingerprinting (including an attempt to distinguish between the 14,000 to 24,000 named cultivars; Thomas and Scott, 1993; Thomas et al, 1993a,b). Collins and Symons (1993) used RAPDs coupled with silver staining of polyacrylamide gels (Bassam et al, 1991) to distinguish several grapevine cultivars.

Fragaria

Commercial strawberry (*Fragaria* x *ananassa*) is an octoploid, and this has hampered its genetic analysis. However, Hancock et al (1994) used RAPDs to investigate the level of polymorphism between eight strawberry cultivars and breeding selections. Their results show that these genetic markers have promise not only for developing a genetic map, but also as a means of verifying cultivars. Davis and co-workers have focused their attention on the closely related diploid strawberry species, *F. vesca* and *F. viridis*, in an effort to genetically define the seven basic linkage groups. They have developed a low resolution map containing 80 markers (i.e., morphological, isozymes, and RAPDs) restricted to four linkage groups. In addition, the mapping suggests the existence of translocation heterozygosity in their interspecific hybrid. Their overall interest is in improving commercial strawberry, through MAS, with respect to tagging genes for fruit color, runner formation, resistance to red stele root rot (Davis et al, 1994), and genetic transformation (Haymes and Davis, 1993).

Vaccinium

A preliminary genetic linkage map with 70 RAPD markers has been constructed using an interspecific hybrid cross of diploid species of blueberry, *Vaccinium* spp. (Rowland and Levi, 1994). The map, intended for use in identifying markers linked to chilling requirements and cold hardiness, covers a genetic distance of 950 cM and comprises 12 linkage groups. Haghighi and Hancock (1992) used RFLP analysis of chloroplast and mitochondrial genomes to investigate the mode of organelle inheritance and to determine the degree of genetic divergence between the four principal sources of cytoplasm in highbush blueberry. The authors demonstrated that mtDNA from all individuals, including inter- and intraspecific hybrids, displayed high degrees of polymorphism enabling them to demonstrate a maternal mode of mtDNA inheritance. In contrast, the cpDNAs from these same plants were identical in DNA banding pattern regardless of enzyme used.

Silver stained gels of RAPDs (Huff and Bara, 1993) were used in an attempt to characterize 22 cranberry (*V. macrocarpon*) varieties (Novy et al, 1994). These markers showed 41% polymorphism and were effective in identifying regional genetic divergence; as well as, establishing unique DNA profiles for 17 of the accessions and in determining the other five to have been misidentified.

Rubus

Similar interests in cultivar identification has spurred work with minisatellite DNA fingerprints and RAPDs in raspberry, *Rubus* spp. (Nybom and Hall, 1991). A recent study demonstrated that 15 cultivars could be distinguished by using only three RAPD primers (Parent et al, 1993).

Vegetables

A comprehensive review of the use of molecular diagnostics in vegetable crops is beyond the scope of this review; however, some specialized applications are worthy of note and are presented below.

Cultivar identification

The identification of and discrimination between distinct vegetable cultivars are essential for the protection of breeder's rights, and to maintain varietal integrity during seed certification. RFLP fingerprints have been used to discriminate between 136 tetraploid potato varieties using four markers (Gorg et al, 1992), and oligonucleotide probes were used to detect highly polymorphic regions in tomato, which can be utilized to identify cultivars. Rus-Kortekaas and Smulders (1992) have demonstrated that the fingerprints produced from tissue culture propagated materials were stable, and did not display differences characteristic of somaclonal variation. RAPDs have been used to characterize several vegetable species including potato (Demeke et al, 1993; Mori et al, 1993), broccoli and cauliflower (Hu and Quiros, 1991), and in celery, where researchers were also able to display results consistent with previously accepted lineages (Yang and Quiros, 1993).

Phylogenetic analysis

An obvious extension of cultivar identification is to utilize DNA diagnostics methods to gain a better understanding of phylogenetic/taxonomic relationships between cultivated crops and their wild relatives. Using the DAF method, Prakash et al (1994) studied the relationships between 103 germplasm accessions of sweetpotato (*Ipomoeae batatas*). The global collection was highly variable, but distinct clustering of accessions was observed based upon geographical source. US cultivars formed a tight cluster and were less diverse than other accessions. Dr. J. Bohac at the USDA/ARS Charleston, SC, USA; (pers. comm.) has been using RFLP, RAPD, and DAF markers to identify the progenitor diploid species of cultivated sweetpotato (6n) as part of the USDA breeding program. These markers will be used for incorporation of desirable genes not present in cultivated species from related wild species through MAS. RFLP analysis has also been used to discriminate between 2n breeding lines and 4n varieties of potato (Gebhardt et al, 1993).

Genome mapping

In contrast to a number of other horticultural crops, the development of a saturated molecular linkage map in vegetables is well underway for many species. These mapping efforts emphasize the tagging of genes for pest resistance or post-harvest characteristics. For example, in lettuce (*Lactuca*), bulked segregant analysis (BSA) was used to identify three RAPD markers that mapped within 12 cM of the $Dm5/8$ gene for downy mildew resistance (Michelmore et al, 1991). BSA is a method where DNA from a segregating F_2 population is pooled together, according to presence or absence of a desired trait, to artificially create near-isogenic lines in species where these are difficult or impossible to create. Furthermore, RAPD markers converted to SCARs allowed for the tagging of downy mildew resistance (Paran and Michelmore, 1993).

A map of the potato (*Solanum tuberosum* ssp. *tuberosum*) genome locates monogenic resistance to the root cyst nematode *(Globodera rostochiensis)*, potato virus X, and *Phytophthora infestans* resistance, and a QTL for *P. infestans* resistance. These RFLP markers for nematode resistance were then converted to allele-specific PCR primers for use in MAS (Gebhardt et al, 1993). A QTL map of potato, tagging genes for tuber dormancy and specific gravity, has also been constructed (Freyre et al, 1994; Freyre and Douches, 1994).

There have been several molecular markers mapped for anthracnose resistance genes in bean (*Phaseolus vulgaris*). Adam-Blondon et al (1994), using near isogenic lines, have identified five RAPD and four RFLP markers linked to within 12 cM of the *Are* gene. Their bean map also contains a dominant gene for male sterility (*Ms8*), and a pod architecture trait (*Sgou*). Similarly, Dirlewanger et al (1994) have constructed a map in pea (*Pisum sativum*) that contains 56 RFLP, four microsatellite and two RAPD loci linked to four disease resistance genes: *Fusarium* wilt, powdery mildew, pea common mosaic virus, and three QTLs for *Ascochyta pisi* blight.

A preliminary linkage map of 34 molecular and morphological markers has been reported (Huestis et al, 1993) in celery (*Apium graveolens*). In this map eight linkage groups are covered with 21 RFLP markers (cDNAs), 11 isozyme loci and two morphological traits.

In tomato, substitution mapping is being used to map QTLs for soluble solids concentration, fruit mass, yield, and pH to very small intervals along a chromosome (Paterson et al, 1990). Substitution mapping is analogous to deletion mapping in humans, where an altered phenotype is associated with the loss of a specific chromosomal region through the use of meiotic recombinants instead of deletions. Giovannoni et al (1991) used pooled F_2 DNA to identify three RAPD markers for opposing alleles for a targeted chromosomal interval, which were defined by linked RFLP markers. These intervals contain genes for regulating pedicle abscission and fruit ripening. This mapping technique can be used to target any chromosomal interval using a single mapping population. More recently, Balint-Kurti et al (1994) performed RFLP analysis with four populations segregating for one of the two closely-linked resistance genes to *Cladosporium fulvum*, and identified 11 tightly linked markers, which are currently being used to clone this gene through positional cloning.

Organellar genome analysis

As mentioned above with fruit crops, characterization of organellar genomes in vegetable crops has proven of basic interest and practical importance (e.g., source of cytoplasm, somatic hybrids, etc.). Identification of S-cytoplasm, the source of cytoplasmic male sterility (CMS) used to produce hybrid-onion seed, has been expedited by the use of molecular markers for cpDNA. Havey (1995) reported a PCR amplification method as a significantly quicker and cheaper alternative to Southern hybridization analysis for identification of S and normal cytoplasms. To examine the mode of inheritance of cpDNA in pea (*Pisum sativum*), RFLP markers were used. Uniparental, maternal inheritance was found, but insufficient F_1 progeny were examined to exclude instances of trace biparentalism (Polans et al, 1990). RAPD markers have also been used in potato to identify somatic hybrids between *Solanum tuberosum* and *S. brevidens* (Xu et al, 1993).

Ornamentals

Genomic analysis of ornamental plants has lagged behind all other crop species. With the exception of petunia and snapdragon, which are used as a model plant systems for gene identification and cloning, very little information is available on the molecular genetics of ornamentals as a group. However, the use of molecular diagnostics in ornamentals have resulted in the development of highly reliable methods for identifying cultivars and assessing the origins of plant hybrids. The most widely used technique is RFLP analysis. Rajapakse et al (1992) used these polymorphisms to fingerprint 16 rose cultivars and found that one probe alone could distinguish 12 cultivars. Hubbard et al (1992) and Hillel et al (1992) also used RFLP profiles to differentiate rose cultivars. Other ornamental cultivars identified by RFLP analysis include gerber daisy and dianthus (Hillel et al, 1992). More recently, the RAPD technique has been used for cultivar identification in chrysanthemum (Wolff and Peters-vanRijn, 1993) and rose (Torres et al, 1993), and to assess the origins of plant hybrids in lilac (Marsolais et al, 1993). Petunia cultivars were distinguished using the DAF technology, allowing conclusions about the phylogeny of commercial lines to be reached (Cerny and Starman, 1994).

Future directions

Future trends in the use of DNA diagnostics in horticultural crops will certainly see the application of these methods expanded to more and more crops. Also, transfer of the more experimental based methods out of the research laboratory and into practical, 'clinical-type' situations will be an important advancement. Technologically, the emphasis is toward automation at all steps, from sample collection and processing to data collection and interpretation. New methods of analysis will be recruited from other disciplines or developed *de novo* (e.g., capillary electrophoresis, chip-based sequencing, genosensors). Regardless of the specific methods, robotics, and improved computational and database capabilities will play a major role in these developments.

In deciduous trees, an efforts to develop a model system for molecular genetic studies and gene isolation is underway. Currently, peach has been singled out for several reasons: 1) the current genetic base is the most extensive; 2) the genome size is very small (approximately twice that of *Arabidopsis*); 3) peach is diploid (n = 8); 4) peach is transformable and can be regenerated from tissue culture; 5) peach can be hybridized with a number of other *Prunus* species, which is useful for genome evolution studies; and 6) the generation time is fairly short (3 to 5 years). A good model species would be invaluable for isolation, cloning, and manipulation of genes important in tree improvement.

Efforts are already well under way to resolve the ambiguity between physical distance and genetic distance in a number of plant systems. This information will in turn dictate the feasibility of map-/marker-based cloning of genes controlling traits for which we lack knowledge of the gene product. A related area of study, which will earn substantial rewards in the field of crop improvement, is that of site-directed (or homologous) recombination. Also, the ability to directly introduce via transformation a molecular identification tag to newly released cultivars will make great strides toward the complete protection of proprietary material and breeders rights.

References

Abbott, A., Eldredge, L., Ballard, R., Baird, V., Callahan, A., Morgans, P. & Scorza, R. (1990) In: *Proceedings SouthEast and National Peach Conf.* (Savannah, GA). pp 23-32.

Abbott, A., Eldredge, L., Ballard, R., Baird, V., Callahan, A., Morgans, P. & Scorza, R. (1991) In: *Proceedings SouthEast and National Peach Conf.* (Hilton Head Island, SC). pp 38-43.

Abbott, A., Belthoff, L., Rajapakse, S., Ballard, R., Baird, V., Monet, R., Scorza, R., Morgens, P. & Callahan, A. (1992) *Plant Genome I*, p 44 (abstr.).

Adam-Blondon, A.F., Sevignac, M., Bannerot, H. & Dron, M. (1994) *Theor. Appl. Genet.* **88**, 865-870.

Akkaya, M., Bhagwat, A. & Cregan, P.B. (1992) *Genetics* **132**, 1131-1139.

Arondel, V.A., Lemieux, B., Hwang, I., Gibson, S., Goodman, H. & Somerville, C.R. (1992) *Science* **28**, 1353-1355.

Arulsekar, S., Parfitt, D., Beres, W. & Hanche, E. (1986a) *J. Hered.* **77**, 49-51.

Arulsekar, S., Parfitt, D. & Kester, D. (1986b) *J. Hered.* **77**, 272-274.

Baird, W.V., Estager, A.S. & Wells, J. (1994) *J. Am. Soc. Hort. Sci.* **119**, 1312-1316.

Balint-Kurti, P. J., Dixon, M.S., Jones, D.A., Norcott, K.A. & Jones, J.D.G. (1994) *Theor. Appl. Genet.* **88**, 691-700.

Bassam, B., Caetano-Anollés, G. & Gresshoff, P.M. (1991) *Anal. Biochem.* **196**, 80-83.

Beckmann, J. & Soller, M. (1991) *Bio/Technology* **8**, 930-932.

Bent, A.F., Kunkel, B.N., Dahlbeck, D., Brown, K.L., Schmidt, R., Giraudat, J., Leung, J. & Staskawicz, B.J. (1994) *Science* **265**, 1856-1860.

Botstein, D., White, R., Skolnick, M. & Davis, R. (1980) *Amer. J. Hum. Genet.* **32**, 314-331.

Bourquin, J.-C., Sonko, A., Otten, L. & Walter, B. (1993) *Theor. Appl. Genet.* **87**, 431-438.

Caetano-Anollés, G., Bassam, B. & Gresshoff, P.M. (1991) *Bio/Technology* **9**, 553-557.

Cai, Q., Guy, C.L. & Moore, G.A. (1994) *Theor. Appl. Genet.* **89**, 606-614.

Cerny, T.A. & Starman, T.W. (1994) *HortScience* **29**, 527.

Chaparro, J.X., Werner, D.J., O'Malley, D. & Sederoff, R.R. (1994) *Theor. Appl. Genet.* **87**, 805-815.

Cheng, F.S., Roose, M.L., Federici, C.T. & Kupper, R.S. (1994) *Plant Genome II* : (abstr. p 32).

Collins, G.G. & Symons, R.H. (1993) *Plant Molecular Biology Reporter* **11**, 105-112.

Condit, R. & Hubble, S. (1991) *Genome* **34**, 66-71.

Damasco, O.P., Godwin, I. D., Henry, R. J., Adkins, S.W. & Smith, M.K. (1994) *Plant Genome III*, p 68.

Davis, T.M., Haymes, K. & Yu, H. (1994) *Plant Genome II*. San Diego, CA. (abstr. #47).

Demeke, T., Kawchuk, L.M. & Lynch, D.R. (1993) *Amer. Potato J.* **70**, 561-570.

Dietrich, W., Katz, H., Lincoln, S., Shin, H.-S., Friedman, J., Dracopoli, N. & Lander, E. (1992) *Genetics* **131**, 423-447.

Dirlewanger, E., Isaac, P. G., Ranade, S., Belajouza, M., Cousin, R. & de Vienne, D. (1994) *Theor. Appl. Genet.* **88**, 17-27.

Durham, R. E., Moore, G.A. & Sherman, W.B. (1987) *J. Amer. Soc. Hort. Sci.* **112**, 1013-1018.

Eldredge, L., Ballard, R., Baird, V., Abbott, A., Morgens, P., Callahan, A., Scorza, R. & Monet, R. (1992) *HortScience* **27**, 160-163.

Foolad, M., Arulsekar, S., Becerra, V., & Bliss, F. (1995) *Theor. Appl. Genet.* **91**, 262-269.

Freyre, R. & Douches, D.S. (1994) *Theor. Appl. Genet.* **87**, 764-772.

Freyre, R., Warnke, S., Sosinski, B. & Douches, D.S. (1994) Theor. Appl. Genet. **89**, 474-480.

Gardiner, S. E., Zhu, J.M., Whitehead, H.C.M. & Madie, C. (1994) *Euphytica* **77**, 77-81.

Gebhardt, C., Mugniery, D., Ritter, E., Salamini, F. & Bonnel, E. (1993) *Theor. Appl. Genet.* **85**, 541-544.

Giovannoni, J.J., Wing, R.A., Ganal, M.W. & Tanksley, S.D. (1991) *Nucl. Acids Res.* **19**, 6553-6558.

Gorg, R., Schachtschabel, U., Ritter, E., Salamini, F. & Gebhardt, C. (1992) *Crop Sci.* **32**, 815-819.

Haghighi, K. & Hancock, J. F. (1992) *HortScience* **27**, 44-47.

Hancock, J. F., Callow, P.W. & Shaw, D.V. (1994) *J. Am. Soc. Hort. Sci.* **119**, 862-864.

Harada, T., Sato, T., Ishikawa, R., Niizeki, M. & Saito, K. (1993) In: *Proc. of the Australia-Japan Workshop on "Techniques of Gene Diagnosis and Breeding in Fruit Trees"*. Hayashi, T., Omura, M. & Scott, N.S. (eds.) Fruit Tree Research Station, Tsukuba, Japan. pp 81-87.

Hashmi, G., Hammerschlag, F., Krusberg, L. & Huettel, R. (1992) *J. Nematol.* **24**, 596.

Hashmi, G. (1993) Ph.D. dissertation. Dept. of Botany, Univ. of Maryland.

Havey, M.J. (1995) *Theor. Appl. Genet.* **90**, 263-268.

Haymes, K. & Davis, T. M. (1993) *Acta Hort.* **388**, 440 (abstr.).

Helentjaris, T., Slocum, M., Wright, S., Schaefer, A. & Nienhuis, J. (1987) *Theor. Appl. Genet.* **72**, 761-769.

Hemmat, M., Weeden, N.F., Maganaris, A.G. & Lawson, D.M. (1994) *J. Heredity* **85**, 4-11.

Hesse, C.O. (1975) In: *Advances in Fruit Breeding.* Janick, J. & Moore, J. N. (eds.), Purdue Univ. Press, West Lafayette, IN. pp 285-335.

Hillel, J., Lavi, U., Tzuri, G. & Vainste, A. (1992) *Acta Hortic.* **314**, 345-351.

Hu, J. & Quiros, C.F. (1991) *Plant Cell Reports* **10**, 505-511.

Hubbard, M., Kelly, J., Rajapakse, S., Abbott, A.G. & Ballard, R.E. (1992) *HortScience* **27**, 172-173.

Huestis, G.M., McGrath, J.M. & Quiros, C.F. (1993) *Theor. Appl. Genet.* **85**, 889-896.

Huff, D.R. & Barra, J.M. (1993) *Theor. Appl. Genet.* **87**, 201-208.

Ishikawa, S., Komori, S., Mikami, T. & Shimamoto, Y. (1993) In: *Proc. of the Australia-Japan Workshop on "Techniques of Gene Diagnosis and Breeding in Fruit Trees".* Hayashi, T., Omura, M. & Scott, N.S. (eds.). Fruit Tree Research Station, Tsukuba, Japan. pp 60-65.

Jelenkovic, G. & Harrington, E. (1972) *Can. J. Genet. Cytol.* 14, 317-324.

Kan, Y. & Dozy, A. (1978) *Lancet* **2**, 910-912.

Landry, B.S., Li, R.Q., Cheung, W.Y. & Granger, R.L. (1994) *Theor. Appl. Genet.* **89,** 847-852.

Laurent, V., Risterucci, A.M. & Lanaud, C. (1993a) *Theor. Appl. Genet.* **87**, 81-88.

Laurent, V., Risterucci, A.M. & Lanaud, C. (1993b) *Heredity* **71**, 96-103.

Laurent, V., Risterucci, A.M. & Lanaud, C. (1994) *Theor. Appl. Genet.* **88**, 193-198.

Lincoln, S., Daly, M. & Lander, E. (1990a) Whitehead Institute Technical Report, 3rd edition.

Lincoln, S., Daly, M. & Lander, E. (1990b) Whitehead Institute Technical Report, 2nd edition.

Litt, M. & Luty, J. (1989) *Amer. J. Hum. Genet.* **44**, 397-401.

Liu, B.H. & Knapp, S.J. (1990) *J. Hered.* **8**, 407.

Marsolais, J.V., Pringle, J.S. & White, B.N. (1993) *Taxon* **42**, 531-537.

Martin, G.B., Brommonschenkel, S., Chunwogse, J., Frary, A., Ganal, M.W., Spivey, R., Wu, T., Earle, E.D. & Tanksley, S.D. (1993) *Science* **262**, 1432-1436.

Matsumoto, S., Tsuchiya, T., Hoshi, N., Soejima, J., Thomas, M., Scott, N. & Ejiri, S. (1993) In: *Proc. of the Australia-Japan Workshop on "Techniques of Gene Diagnosis and Breeding in Fruit Trees".* Hayashi, T., Omura, M. & Scott, N.S. (eds.) Fruit Tree Research Station, Tsukuba, Japan. pp 26-30.

Matsuyama, T., Omura, M. & Akihama, T. (1993) In: *Proc. of the Australia-Japan Workshop on "Techniques of Gene Diagnosis and Breeding in Fruit Trees".* Hayashi, T., Omura, M. & Scott, N.S. eds.) Fruit Tree Research Station, Tsukuba, Japan. pp 14-25.

Meksem, K., Leister, D., Peleman, J., Zabeau, M., Salamini, F & Gebhardt, C. (1995) *Mol. Gen. Genet.* **249**, 74-81.

Messeguer, R., Viruel, M., de Vicente, M., Garcia-Mas, J., Fernandez-Busquets, X., Vargas, F., Puigdomenech, P. & Arus, P. (1994) *Plant Genome II*, 51 (abstr. P145).

Michelmore, R.W., Paran, I. & Kesseli, R.V. (1991) *Proc. Natl. Acad. Sci. USA* **88**, 9828-9832.

Mindrinos, M., Katagiri, F., Yu, G.-L. & Ausubel, F.M. (1994) *Cell* **78**, 1089-1099.

Monet, R. (1989) *Acta Hortic.* **254**, 49-57.

Monet R., Bastard, Y. & Gibault, B. (1985) *Agronomie* **5**, 727-731.

Morgante, M. & Olivieri, A.M. (1993) *Plant J.* **3**, 175-182.

Mori, M., Hosaka, K., Umemura, Y. & Kaneda, C. (1993) *Jap. Jour. Genet.* **68**, 167-174.

Mowrey, B.D., Werner, D. J. & Byrne, D. H. (1990) *J. Amer. Soc. Hort. Sci.* **115**, 312-319.

Mullis, K.B., Faloona, F., Scharf, S., Saiki, R., Horn, G. & Erlich, H. (1986) *Cold Spring Harbor Symp. Quant. Biol.* **51**, 263-273.

Novy, R., Kobak, C., Goffreda, J. & Vorsa, N. (1994) *Theor. Appl. Genet.* **88**, 1004-1010.

Nybom, H. & Hall, H. (1991) *Euphytica* **53**, 107-114.

Nybom, H., Rogstad, S. & Schall, B. (1990) *Theor. Appl. Genet.* **79**, 153-156.

Olson, M., Hood, L.R., Cantor, C. & Botstein, D. (1989) *Science* **245**, 1434-1435.

Omura, M., Hidaka, T., Nesumi, H., Yoshida, T. & Nakamura, I. (1993) In: *Proc. of the Australia-Japan Workshop on "Techniques of Gene Diagnosis and Breeding in Fruit Trees".* Hayashi, T., Omura, M. & Scott, N.S. (eds.) Fruit Tree Research Station, Tsukuba, Japan. pp 66- 73.

Ouazzani, N., Lumaret, R., Villemur, P. & Di Giusto, F. (1993) *J. Heredity* **84**, 34-42.

Paran, I. & Michelmore, R.W. (1993) *Theor. Appl. Genet.* **85**, 985-993.

Parent, J.-G., Fortin, M. & Page, D. (1993) *Can. J. Plant Sci.* **73**, 1115-1122.

Paterson, A.H., DeVerna, J.W., Lanini, B. & Tanksley, S.D. (1990) *Genetics* **124**, 735-742.

Polans, N.O., Corriveau, J. & Coleman, A. (1990) *Curr. Genetics* **18**, 477-480.

Pontikis, C.A., Loukas, M. & Kousounis, G. (1980) *J. of Hort. Sci.* **55**, 333-343.

Prakash, C.S., He, G. & Jarret, R.E. (1994) *HortScience* **29**, 727.

Quarta, R., Dettori, M.T., Verde, I., Laino, P., Santucci, F., Vantaggi, A., Sabatini, S. & Sciarroni, R. (1994) Convegno Agro biofruit su "Techologie avanzate per l' identificazione varietale e il controllo santario nel vivaismo fruttiviticolo".

Rajapakse, S., Belthoff, L.E., He, G., Estager, A. E., Scorza, R., Verde, I., Ballard, R.E., Baird, W.V., Callahan, A., Monet, R. & Abbott, A.G. (1995) *Theor. Appl. Genet.* **90**, 503-510.

Rajapakse, S., Hubbard, M., Kelly, J.W., Abbott, A.G. & Ballard, R.E. (1992) *Scientia Hortic.* **52**, 237-245.

Ritter, E., Gebhardt, C. & Salamini, F. (1990) *Genetics* **125**, 645-654.

Rogstad, S., Patton II, J. & Schall, B. (1988) *Proc. Natl. Acad. Sci. USA* **85**, 9176-9178.

Roth, J. & Lark, K. (1984) *Theor. Appl. Genet.* **68**, 421-431.

Rowland, J. & Levi, A. (1994) *Theor. Appl. Genet.* **87**, 863-868.

Rus-Kortekaas, W. & Smulders, M.J. (1992) *Theor. Appl. Genet.* **85**, 239-244.

Schnell, R.J. & Knight, R.J. (1993) *Acta Hortic.* **341**, 86-91

Scorza, R., Ravelonardo, M., Callahan, A.M., Cordts, J., Fuchs, M., Dunez, J. & Gonsalves, D. (1994) *Plant Cell Reports* **14**, 18-22.

Sondur, S., Manshardt, R. & Stiles, J. (1996) *Theor. Appl. Genet.* **93**, 547-553.

Stam, P. (1993) *Plant J.* **3**, 739-744.

Suiter, K.A., Wendel, J.F. & Case, J.S. (1983) *J. Heredity* **74**, 203-204.

Tanksley, S.D. (1983) *Plant Mol. Biology Reporter* **1**, 3-8.

Tanksley, S.D., Young, N.D., Paterson, A.H. & Bonierbale, M.W. (1989) *Bio/Technology* **17**, 257-264.

Tautz, D. (1989) *Nucleic Acids Research* **17**, 6463-6471.

Thomas, M.R. & Scott, N.S. (1993) *Theor. Appl. Genet.* **86**, 985-990.

Thomas, M. R., Matsumota, S., Cain, P. & Scott, N. S. (1993a) *Theor. Appl. Genet.* **86**, 173-180.

Thomas, M.R., Cain, P., Matsumoto, S. & Scott, N.S. (1993b) In: *Techniques of Gene Diagnosis and Breeding in Fruit Trees.* Hayashi, T., Omura, M. & Scott, N.S. (eds.) Fruit Tree Research Station, Tsukuba, Ibaraki, Japan. pp 7-13.

Torres, A. M., Millán, T. & Cubero, J.I. (1993) *HortScience* **28**, 333-334.

Vallejos, C. E., Sakiyama, N.S. & Chase, C.D. (1992) *Genetics* **131**, 733-740.

Vos, P., Hogers, R., Reijans, M., van de Lee, T., Hornes, M., Frijters, A., Pot, J., Peleman, J., Kuiper, M. & Zabeau, M. (1995) *Nucl. Acids Res.* **23**, 4407-4414.

Weber, J. L. & May, P.E. (1989) *Amer. J. Hum. Genet.* **44**, 388-396.

Weissenbach, J., Gyapay, G., Dib, C., Vignal, A., Morissette, J., Millasseau, P., Vaysseix, G. & Lathrop, M. (1992) *Nature* **359**, 794-801.

Welsh, J. & McClelland, M. (1990) *Nucleic Acids Research* **18**, 7213-7218.

Williams, J., Kubelik, A., Livak, K., Rafalski, J.A. & Tingey, S. (1990) *Nucleic Acids Research* **18**, 6531-6535.

Wolff, K. & Peters-vanRijn, R. (1993) *Heredity* **71**, 335-341.

Wyman, A. & White, R. (1980) *Proc. Natl. Acad. Sci. USA* **77**, 6754-6758.

Xu, Y.-S., Clark, M.S. & Pehu, E. (1993) *Plant Cell Reports* **12**, 107-109.

Yamamoto, M., Kobayashi, S., Nakamura, Y. & Yamada, Y. (1993) In: *Proc. of the Australia-Japan Workshop on "Techniques of Gene Diagnosis and Breeding in Fruit Trees"*. Hayashi, T., Omura, M. & Scott, N.S. (eds.) Fruit Tree Research Station, Tsukuba, Japan. pp 39- 46.

Yang, X. & Quiros, C.F. (1993) *Theor. Appl. Genet.* **86**, 205-212.

Ye, X., Brown, S., Scorza, R., Cordts, J. & Sanford, J. (1994) *J. Amer. Soc. Hort. Sci.* **119**, 367-373.

Young, N. D. & Tanksley, S.D. (1989) *Theor. Appl. Genet.* **77**, 353-359.

Zhao, X. & Kochert, G. (1993) *Plant Mol. Biology* **21**, 607-614.

Commercial Applications of DNA Profiling by Amplification with Arbitrary Oligonucleotide Primers

Peter M. Gresshoff and Gustavo Caetano-Anollés

Plant Molecular Genetics, Center for Legume Research and Institute of Agriculture
The University of Tennessee , Knoxville, Tennessee 37901-1071, USA

Introduction

The development of DNA analysis methods has provided a new opportunity for technology transfer. There are numerous commercial applications of the technology. These extend from forensic uses such as those seen for the O.J. Simpson trial, or DNA analyses for turfgrass growers and customers as illustrated here. Pet as well as race horse owners are interested in DNA analysis to verify pedigrees. The patent office requires often additional information which can be obtained through DNA fingerprinting.

The ultimate form of genetic analysis and therefore DNA profiling, is the complete sequencing of an organism and the comparison to that of a second. This is unrealistic. Hence, smaller regions of the genome need to be compared. This was originally done through the use of genetic probes such as those used in RFLP technology. Over time more sensitive methods, often based on the PCR (polymerase chain reaction) found application. This chapter outlines some of our efforts in the commercial application of the DNA amplification fingerprinting (DAF) technology.

The need for DNA profiling

Molecular genetic approaches have enriched the resolution of plant genome analysis. The ability to clone and sequence specific regions of almost any genome has provided sequence-based information, which adds to our understanding derived from cytogenetics and large scale DNA analyses, such as those from reassociation kinetics and flow cytometry.

While databases of DNA sequences are exponentially growing, further methods are needed to investigate plant genomes at a level of complexity **above** the primary sequence, but **below** the cytogenetic, karyotypic arrangement.

The same requirements exist for taxonomic and population genetic studies. Convenient, low cost, but reliable methods are needed to categorize seed collections and natural populations with the purpose of optimizing resource management and preservation. Numerous molecular tools seem to be available ranging from costly DNA sequencing of localized regions to RFLP and microsatellite analysis.

Many of these molecular techniques are specific for a species or genus of choice. Expansion to a new species requires a substantial amount of ground work, thereby limiting application and extending costs. For example, a set of microsatellite primers designed for the human genome may only have marginal use in the analysis of bovine DNA, but definitely would fail with soybean DNA.

General molecular diagnostic techniques, however, were developed in recent years. Essentially, these are based on arbitrary oligonucleotide primers, which explore multiple regions of a genome in a PCR-like manner. Differences in the pattern of amplified DNA are used to determine the degree of similarity between sample organisms.

This chapter will focus on one of these techniques, namely the DAF method developed and patented (US Patent Number 5,413,909) in our laboratory, and will feature some of its applications in commercial diagnostics, operated through the laboratory at the University of Tennessee.

DNA profiling using DAF

Arbitrary primer-based DNA amplification techniques extend the utility of the Polymerase Chain Reaction (PCR) to general genome analysis (Figure 1). MAAP procedures were developed independently, and apparently concurrently, in three laboratories (DAF, RAPD and AP-PCR; Caetano-Anollés et al, 1991a; Williams et al, 1990; Welsh and McClelland, 1990). Two initial patents, issued in the USA, cover the original strategy for the RAPD and DAF version. Recently, a Dutch company called KeyGene developed another high resolution fingerprinting technique, called SRFA (Selected Restriction Fragment Amplification), in which restriction fragments are selectively amplified from ligated linker sites (Smith et al, 1993; Vos et al, 1995). Because of a plethora of terms we proposed the general and unifying acronym MAAP (**M**ultiple

Arbitrary Amplicon Profiling; Caetano-Anollés et al, 1992b, 1993) to describe the general strategy of using single arbitrary primers to profile genomes.

---

identity testing for legal cases
gene beacons/molecular markers for positional cloning
preliminary exclusionary testing for paternity cases
trait introgression
backcross conversion
map saturation for marker assisted selection (MAS)
generation of sequence tagged sites on YACs
fingerprinting and STS development of PCR products
determination of population structure for taxonomy and ecology
pathogen identification (especially non-cultureable mycoplasms)
genetic identification for plant variety registration

Table 1: Some applications of DNA amplification fingerprinting

In essence, MAAP involves the use of short, arbitrarily chosen oligonucleotide primers, which annealed to DNA, will direct DNA amplification of multiple genome regions (amplicons; Mullis, 1991). Temperature cycling and the use of a thermostable DNA polymerase are common components with the more specific and targeted PCR. In contrast to PCR, MAAP procedures generally use a single primer of arbitrary sequence. Some primers may be structured, as are the mini-hairpin primers recently used by Caetano-Anollés and Gresshoff (1994a) to fingerprint a wide variety of genomes and DNA fragments. Prior sequence information, as needed for PCR, is not required. MAAP intentionally generates multiple products, which itself would be a rather undesirable result in a PCR reaction. MAAP is universal, so that a primer used for one species can be used repeatedly for others, even if evolutionary distances and DNA complexities between the template nucleic acids are significant.

Amplification products are separated and recorded by a variety of detection methods; in all cases, a linear array of signals generates a profile, which is representative for the target DNA and specified by the DNA sequence of the primer. Variations in primer sites on the target DNA, length variations between primer sites, and possibly changes in the secondary structure of target DNA between or flanking the primer recognition sites generate molecular polymorphisms. These amplification polymorphisms define molecular regions of the plant genome and thus can be used as (i) potential sequence tagged sites for positional cloning approaches, or (ii) components of profile used in DNA fingerprinting and diagnostics.

Amplification polymorphisms were termed AFLP (amplification fragment length polymorphism) by analogy to RFLP (Caetano-Anollés et al, 1991a,b). Recently, the same term has been applied to polymorphisms generated by the AFLP method, involving DNA digestion, adapter ligation, and primer amplification (Vos et al, 1995; Smith et al, 1993). It may be acceptable to use the same term for the final recorded polymorphism, even if the underlying mechanism used in the generation was different. However, the term itself may be slightly misleading. Genetic studies on MAAP markers showed that the majority are inherited as dominant markers (see Prabhu and Gresshoff, 1994; Williams et al, 1990). Polymorphisms are based on either the loss of an amplification site, or a molecular change, which lowers the efficiency of amplification for the one allele. Such molecular changes may be alterations in the primer site, but also may involve length variation, leading to negative selection during amplification. Recalling that DAF is based on an exponential amplification paradigm, small amplification disadvantages, possibly caused by fold-back DNA and secondary structure, will result in an apparent absense of a band. Hence, it is likely that most polymorphisms are not caused by length variations but by primer inefficiencies.

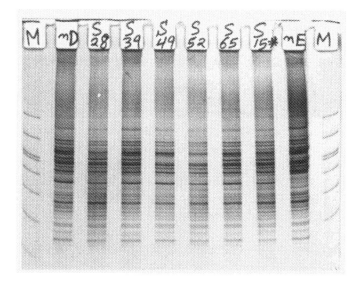

Figure 1: DNA amplification fingerprint gel derived from different soybean lines. Primer HpD-2 (a mini-hairpin primer) was used for the 8 samples. nD and nE are parental soybean genotypes while lanes labels 'S' are F$_2$ segregants. A molecular polymorphism at about 380 bp segregates 100 % in the F$_2$. Molecular markers range from 1000 bp to 100 bp. Gel courtesy of Prof. Bahman Yazdi-Samadi (The University of Tennessee, Knoxville).

Caution is needed to prevent experimental variability, non-parental bands, and artifacts. RAPD parameters, such as very high template DNA concentration (often as high as 5 ng per microliter of reaction mixture) and low primer concentration (usually around 0.3 µM), increase variability as frequently seen in published figures (including the seminal paper by Williams et al, 1990). RAPD patterns from known pedigrees showed non-parental bands appearing in the F_1. These may be artifacts caused by mismatching, variable DNA quality, recombination between amplification products caused by reannealing of internally repeated DNA segments commonly found in eukaryotic DNA, or secondary products generated on the first generation products.

Furthermore, we believe that not all amplicons defined by a primer pair are equally represented in the final amplification products (Caetano-Anollés, 1994). This may cause stochastic amplification results as some reactions may be favored in the first two cycles under some conditions and not others. It is possible that competition for primer occurs during cycling. Secondary amplification may occur, as DAF products arising during early rounds of amplification serve as template for secondary primer annealing sites. Repeated regions inside DNA products may reanneal prior to primer annealing causing lower efficiency amplification and possible loss from detection in an ethidium bromide stained agarose gel (Caetano-Anollés et al, 1992c). While such artifacts may occur in all amplifications, including PCR and DAF, they appear more likely in the RAPD procedure because of unfavorable amplification parameters such as high input DNA levels and low primer concentration.

Of all MAAP procedures, DAF utilizes the shortest primers, down to 5 nucleotides in length. The optimal length was found to be eight nucleotides, a length which does not produce efficient amplification with RAPD amplification conditions. Informative profiles were generated with five-nucleotide primers (5-mers) using soybean DNA as a template (Caetano-Anollés et al, 1993). DAF products are routinely separated by thin polyacrylamide gels, backed onto plastic Gel-Bond film run in a commercially available vertical gel chamber (such as the Mini-Protean II unit; BioRad. Inc., CA). This gel-plastic support provides backing during subsequent washing steps and helps preserve the original gel. The gel is stained by an improved silver staining method, which detects DNA at about 1 pg·mm^{-2} (Bassam et al, 1991; Caetano-Anollés and Gresshoff, 1994b). This silver staining technology is patented (US Patent Number 5,643,479) and is marketed by Promega Inc. as their SilverSequence™ kit, useful for staining DNA sequencing gels, microsatellites, differential display gels as well as DAF amplifications.

Special care is needed to obtain high contrast/low background silver staining by using high quality chemicals, fresh staining solutions, and avoiding formaldehyde which has been stored below 4°C or is of old age (Caetano-Anollés and Gresshoff, 1994b). Resultant gels are air-dried and kept for permanent record and evaluation. A pattern of bands, reminiscent of the Universal Product Code (UPC) is generated, permitting the easy distinction of genotypes (Figure 1). The silver staining procedure has found additional utility in other applications, such as silver staining of DNA sequencing gels (Caetano-Anollés and Gresshoff, 1994b; Kruchinina and Gresshoff, 1994) or the detection of microsatellite (Gordon Lark, [Utah]; Lee Roy Hood [Seattle], pers. comm.). Indeed, Promega Corp. markets the silver staining method of Bassam et al (1991) through the SilverSequence® line of products under license from the University of Tennessee.

Pattern detection

DAF generates multiple amplification products which can be visualized on silver stained PAGE gels as bands of different intensity. Theoretical analysis showed that in complex genomes, such as soybean, not all existing amplicons are amplified effectively enough to be scored by PAGE and silver staining (Caetano-Anollés, 1994). In general, DAF profiles contain about 20 to 40 scoreable bands. This number varies with primer and template DNA, and can be increased considerably using DNA separation techniques with higher band resolution. Comigration of bands is likely on shorter gels, therefore requiring cloning and sequencing to generate verified unique molecular markers (Weaver et al, 1994; Kolchinsky et al, 1997).

PAGE/silver staining provides low cost, high through-put analysis of DAF products. Profiles are easily stored in photo albums, without the need for photography. The gel itself offers a depository of the actual DNA, facilitating subsequent analysis or cloning of DNA fragments (Weaver et al, 1994; Kolchinsky et al, 1993). PAGE gels dry easily in air and can be scanned by laser densitometry (see Gresshoff and MacKenzie, 1994; Prabhu and Gresshoff, 1994).

We have found that some amplification patterns are too complex to be resolved on the Mini-Protean gels; for this reason we ran long (40 to 60 cm) DNA sequencing type PAGE gels and stained these with silver (Shane Abbitt, UT Knoxville, pers. comm.; Kolchinsky et al, 1997). Such gels provide high resolution (as many as 100 bands can be detected) and relatively fast sample throughput, although the expenditure on chemicals and effort is increased.

DAF products were also resolved by alternative methods. Agarose gels give clear resolution, but fewer products (Prabhu and Gresshoff, 1994; Caetano-Anollés, 1994). For example, Kolchinsky et al (1993) amplified isolated yeast chromosomes with DAF primers, separated the products on agarose, isolated individual bands by cloning, and re-hybridized the cloned fragments onto pulse field gel (PFG) karyotypes of yeast. Cloned fragments detected only their appropriate template chromosome.

Similar cloning, but following reamplification prior to cloning of picked bands from dried, silver stained polyacrylamide gels was achieved by Weaver et al (1994). Many DAF products were sequenced in our laboratory, revealing the primer structure at both ends of the correctly sized product. Cloned DAF bands of the same size may contain two different DNA sequences, suggesting co-migration (R. Prabhu, UT Knoxville, pers. comm.).

In collaboration with Applied Biosystems Inc. (Foster City, CA) fluorochrome labeled primers were used to direct amplification of plant DNA (Caetano-Anollés et al, 1992a). The resultant amplification products were separated on an ABI Sequencer using Gene Scanner software. Similar attempts at automated separation using capillary electrophoresis (CE) have been promising (Caetano-Anollés et al, 1995), providing separation of single samples in 30 minutes. Additionally, modern CE instruments have autosamplers and provide multi-channel (as much as 96 channels) separation. This may allow significant labor and cost reduction through automation.

In general, DAF generates scoreable polymorphisms in the molecular size range from 100 to 800 bp. Recently, we have used the pre-cast and automated PhastGel system (Pharmacia Inc.) to obtain profiles for pathogenic nematodes on soybean (Baum et al, 1994). The precast gradient PAGE gels and the automated silver staining permitted rapid throughput (160 samples per day). Bands at higher molecular weight (up to 1,500 bp) were scoreable; species and race-specific polymorphisms were detected. Denaturing gradient gel electrophoresis (DGGE) is another method which would help to distinguish polymorphic products of wheat (He et al, 1992).

In summary, detection can be by several procedures, offering different costs, convenience, and resolution. Automation is possible at present but may require lower cost machines and procedures to make MAAP of great utility to the plant breeder or molecular geneticist.

In this context it is noteworthy that DuPont Company has developed a DNA isolation machine (Rafalski et al, 1994). This machine is now being housed in a new DNA plant breeding facility in Iowa, close to the field sites, where breeders require marker-assisted breeding of corn and soybean.

Genetic uses of DAF

The ability to detect molecular markers closely associated with genes of agricultural importance, makes marker-based breeding an attractive proposition. The need for maintaining large plant populations through advanced breeding cycles can be reduced by detecting heterozygotes. MAAP markers converted through cloning, partial sequence analysis and specific PCR primer synthesis may provide SCARs (sequence characterized amplified regions; Paran and

Michelmore, 1993), which are diagnostic for either a gene region in a plant or a pathogen.

Figure 2 cartoons the utility of RFLPs and MAAP markers in generating diagnostic tools. Closely linked molecular markers, either defined by an RFLP or an amplification polymorphism (AP), can be sequenced to define specific PCR primers. PCR tests are more specific and easier to score. This strategy was successfully used by Kolchinsky et al (1997) for both RFLP and DAF markers closely linked to the supernodulation locus of soybean.

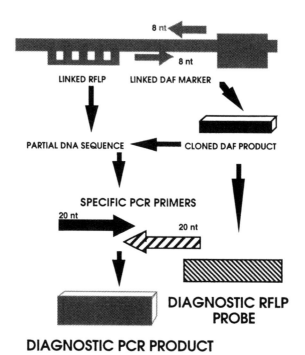

Figure 2: RFLP and MAAP markers used as diagnostic tools for genome analysis. Partial sequencing of clones, which are linked to your favorite gene (yfg1) provides information for specific PCR primers, which in turn generate a diagnostic product. A sequence characterized amplified region (SCAR) was demonstrated for the supernodulation gene of soybean by Kolchinsky et al (1997).

An RFLP marker, closely linked in F_2 segregation to the supernodulating gene of soybean (Landau-Ellis et al, 1991), was sequenced and PCR primers generated which detect the polymorphism between the parental genotypes. The PCR test allowed quick and easy analysis of large populations. Likewise, a DAF polymorphism was converted into a SCAR. A sequence of a polymorphic DAF product detected by bulk segregant analysis of DNA pools distinguished by their source from supernodulating or wild-type soybean plants allowed the synthesis of specific PCR primers, which gave a polymorphic product (Kolchinsky et al, 1995). This SCAR was mapped to the region of the *nts* locus on Molecular Linkage Group H of the soybean RFLP map (Shoemaker and Olson, 1993). Interestingly, Kolchinsky et al (1997) used sequencing length PAGE gels and autoradiography to reveal the linked polymorphisms. The same primers were used to perform *in situ* PCR on soybean chromosomes and detected a single copy amplification product (Paul Keim, Univ. Northern Arizona, pers. comm.). This illustrates how detailed molecular information about a gene-associated sequence (or even better the gene itself) helps to further science in several disciplines.

It may be possible to find a marker specific for a soybean nematode race (see Baum et al, 1994), to convert it to a SCAR, then use a diagnostic, proactive test on agricultural soil to predict, which nematode race is predominant in the field prior to planting. In a similar fashion, Tamut and Gresshoff (unpubl. data) developed a diagnostic DNA test for citrus greening disease. This disease affects citrus in Nepal and neighboring countries. The pathogen has not been identified, and only symptoms permit the farmer to detect the disease. At this stage, significant time and space have been used. Early detection at the nursery stage would facilitate better control. Diseased and healthy plants of the same genotype were used. Although plant genotypes differed, a common band was detected in diseased material. This band was cloned, sequenced and used as a diagnostic tool. Data bank searches suggest sequence similarity to a mycobacterium or a virus (B. Tamut, UT Knoxville, unpubl. data).

Rafalski et al (1994) described the use of automated machines that isolate DNA from plant tissue and robotics that help separate amplified DNA and interpret DNA profiles. Plant genome analysis for agricultural purposes, as compared to medical or forensic diagnoses in humans, is different in many cost and volume parameters. Plant analysis generates high numbers of samples, which need to be analyzed at an intermediate resolution and accuracy. Accidentally destroying a recombinant plant has different implications than imprisoning a person for 20 years! Costs per analysis need to be low in plants because the final product is of relatively low unit value, and large numbers of samples need to be tested to achieve the marketable product. Additionally, both socially and ethically, our society has (correctly or not) always placed a lower monetary value on plant research (Gresshoff, 1990).

DAF markers were shown to be repeatable polymorphisms in different DNA isolations, operators, time periods, and amplifications. They are heritable, as about 75% of AFLPs between *G. max* and *G. soja* segregated as dominant Mendelian markers in F_2 populations (Prabhu and Gresshoff, 1994; Caetano-Anollés et al, 1993). Interestingly, the other 25% segregated in an uniparental way, being either maternal or paternal or non-parental. Maternal inheritance presumably stems from amplification of cytoplasmic replicons.

Several DAF polymorphisms were mapped in recombinant inbred lines of soybean (Prabhu and Gresshoff, 1994). The use of inbred lines is very convenient for DAF, as the lines are predominantly homozygous (Lark et al, 1993). Since DAF markers are seemingly dominant, it is impossible to distinguish the dominant homozygote from the heterozygote. Accordingly, in normal F_2 populations, larger sample numbers are required to obtain data equivalent to those obtained from the analysis of a codominant (e.g., RFLP) marker. In recombinant inbreds, however, DAF and RFLP markers share the same statistical advantages.

MAAP markers are useful in defining closely linked regions in bulked segregant analysis (Michelmore et al, 1991; Kolchinsky et al, 1997). The availability of large primer sets and the generation of multiple amplification products results in the efficient screening of the genome. However, not all detected polymorphisms between bulks are reliable and needed to be confirmed by repeated amplification. It is not uncommon to hear of researchers, who tested 1,000 single primers to find 10 or 12 polymorphisms between bulked DNA samples or near-isogenic lines (NILs), only to find that one or two map close to the locus of interest.

Caetano-Anollés and Gresshoff (1994b) used mini-hairpin primers in a DAF reaction to profile soybean YACs. The mini-hairpin primers are interesting, because they contain on their 5' end a 7 nucleotide fold-back loop (4 nt in stem, 3 nt in the loop). The 3' end can be as short as 3 nt, allowing the generation of a small set of 64 primers, which are useful for the characterization especially of small genomes or genome components such as plasmids or YACs.

These findings show that single primer DNA amplification analysis of plant genomes adds a further genetic tool to construct high density maps needed for positional cloning and marker-based breeding approaches. The large number of products allows a high density genotyping and genotype differentiation (Gresshoff, 1992). Reliable exclusion is obtained when one or more bands differ between samples. Inclusion is more difficult, as many primers need to be tested, frequency of variation within the sampled species needs to be known, and careful statistical statements need to be generated. It is impossible to declare with 100% certainty that two things are the same; it always needs to be a probabilistic statement. It is up to the user (society, courts, scientists) to concur on acceptable levels of confidence for such probabilities.

DAF allowed the easy distinction of variant turfgrass material in commercial plots (Callahan et al, 1993). The application of DNA tests to the turfgrass industry is a major challenge in an area of repeated vegetative propagation, triploidy, and genetic instability. For example, foundation stock from several geographic locations gave identical profiles for bermudagrass cultivar Tifway, while samples analyzed from golf course owners repeatedly showed major variation. Indeed, one supplier gave us samples over a one year period and the same off-variant of Tifway was detected. We found DAF primers which distinguish between Tifway and Tifway II. The latter is a radiation-induced variant. Likewise, it is possible to separate Tifgreen and the "spontaneous" mutant Tifdwarf (Caetano-Anollés et al, 1995). DAF also allowed a detailed phylogenetic analysis of established centipedegrass and bermudagrass cultivars (Weaver et al, 1995; Caetano-Anollés et al, 1995).

Sunflower material provided by a seed company was categorized into several groups. Some common bands permitted the suggestion of a possible breeding pedigree and close relatedness. This type of analysis has utility for product verification and plant variety rights. DAF analysis may also reveal the putative parents of a commercially protected hybrid (say maize or sunflower), permitting the reconstruction of the hybrid by renewed breeding from unprotected parental stock.

DAF was used to analyze the relationship between 10 commercial soybean genotypes. These lines were closely related at times, representing backcross material as well as direct hybrids. Seven DAF primers recognized genotype differences which allowed the construction of a phylogenetic tree (Prabhu et al, 1997). Major groupings and relative distances were in close agreement to those discovered in a separate study on the same genotypes using 53 RFLP probes and the breeding pedigree tree (Prabhu et al, 1997). These findings provide strong support for the utility of DAF markers for commercial soybean genotype recognition.

The determination of genetic identity is not only important to the seed manufacturer, but also for the determination of plant product quality as many food manufacturers use processes directly optimized for a specific biological feedstock. This industry relies on biological material; it is essential that quality biological feedstock enters the manufacturing process. Often it is impossible to inspect the source plant, as one looks at a harvested product. It is for these industrial and related horticultural applications that a new technology was needed. The advent of DNA analysis has provided an additional way by which closely related organisms are distinguished for industrial, manufacturing, and retailing purposes.

How are samples to be sampled and deposited? Which techniques are acceptable? Who will (if at all) certify laboratories? What controls need to run on each gel? Sofar the human forensic, paternity, and criminological applications of DNA profiling have received much legal and commercial interest. Plant genotyping for agricultural, legal, and ecological purposes is still in its infancy. It appears as if international and national guidelines are needed.

The need for commercial plant DNA analysis laboratories

Our research has laid the basis for the scientific application of the DAF technology. We have focused on understanding the genetic and molecular uses. Significant advances were achieved. For several years the researchers at our laboratory have conducted DNA identity tests for outside parties. Most of these dealt with turfgrass, because there are few morphological markers which can be used reliably to distinguish variant grass lines. Our laboratory has been recommended by the Florida branch of the USGA for turfgrass DNA analysis.

We have established mechanisms of efficient sample and data handling. Reports are generated for clients, permitting them to market genetically tested material. Alternatively, clients have used DAF data to confirm the identity of their own product over that of a competitor. Table 2 lists the plant species, which have been successfully fingerprinted with the DAF method.

Over the last 3 years our laboratory has successfully fingerprinted 300 commercial samples. This interest was generated by 'word-of-mouth'. No advertisement was placed as our laboratory has only limited resources to handle large sample volume. We have achieved a high level of customer satisfaction, with samples having been received from several sources on repeated occasions. Clients come from as far as the Malaysia, Australia, the Philippines and Indonesia. While the major clients are golf club managers, we have sampled turfgrass for the professional football stadiums in Dallas, Knoxville, and Miami.

Several major seed companies have sent samples to evaluate plant material. This includes genetically engineered tomato from Calgene Inc. (see chapter in this volume), sunflowers from Cargill Seed company and soybean from Pioneer HiBred . Several minor sod farms routinely send material for verification prior to sale. This provides insurance and increases the product value as he customer is assured of genetic quality control.

Based on these successes, it appears clear that while this type of research can be inititated in a university laboratory, and intermittent service can be provided to the public on a semi-commercial (not-for-profit) way, the proper commercialization should be carried out in an independent laboratory. Best results may be obtained if the commercial laboratory can draw on experience and temporary assistance from the university laboratory and the inventors. The interaction of researchers and commercial staff can be assured through proximity and interactive appointments. Basic researchers can funnel information on new technologies to the company. In return the company can sponsor project work in the laboratory and provide nonacademic training for graduate students.

plant species	plant species
field crops:	*horticultural crops:*
soybean	nandina
cotton	petunia
peanut	tomato
sunflower	*woody plants:*
sweet potato	dogwood
	white pine
turfgrasses:	oaks
Zoysiagrass	*model plants and others:*
centipedegrass	Azolla
Kentucky bluegrass	*Lotus japonicus*
St. Augustinegrass	cannabis
bermudagrass	*Medicago truncatula*

Table 2: Plant species that have been successfully fingerprinted using the DAF method. Peanut and sweet potato were done by Dr. Prakash (Tuskegee University, Alabama). Petunia was done by Dr. T. Starman and Ms. T. Cerny L. *japonicus and* M. truncatula *are model legume species of little agronomic value.*

Commercial DNA testing laboratories would also provide an employment opportunity for graduate students. They would learn (on a part-time basis) the commercial realities of applied science. This combination of scientific with business and management skills would increase their future employability.

Investment is needed for licensing and commercial development. Equipment and staff costs are expected to be substantial in the beginning stages, but any company has the potential of immediate cash-flow, as well as an established market niche. Automation in DNA extraction and separation would drive down costs per sample, thereby increasing the utility to a broader market.

Commercial laboratories could broaden the horizons of application. For example, the human paternity testing area could be helped through DAF by preliminary exclusionary testing (PET). Nearly one third of cases are exclusions (R. Bever, Genetic Designs, North Carolina, pers. comm.). The cost of detailed VNTR or RFLP analysis could be spared, if excluded samples can be detected by a low cost method such as DAF. Obviously, these laboratories could engage in new product development such as the manufacture of diagnostic kits, complete with primers, standards, buffers and instructions. Using immobilized oligonucleotides, synthesized rapidly by lithographic methods, separated on membranes or plastic supports, one could envision the development of new Profiling-by-Hybridization (PBH) technologies. With such technologies, sample DNA would bind to a solid support and could be detected electronically in a dot pattern, perhaps even of different densities. Separation technologies would no longer be required, as the oligonucleotides are pre-separated. Such technology would go hand-in-hand with the developing Sequencing-by-Hybridization (SBH) approaches.

Cost of DNA profiling

A major consideration for market acceptance of a new technology is the utility and need of the technology, the quality of the end product, and the price. At present DNA profiling is expensive. If radioactive probes are used, the fingerprinting technique becomes too laborious and limited to small sample sizes in specialty cases. If PCR is used, then the commercial source of the enzyme and the manual input for DNA isolation become a major factor.

material	cost per sample in US$		
	DAF	AP-PCR	RAPD
Amplification:			
enzyme	0.640	0.640	0.640
primer	0.012	0.030	0.020
reactants	0.015	0.015	0.015
$[\alpha\text{-}^{32}P]dATP$	-	0.198	-
disposables	0.150	0.150	0.140
DNA resolution:			
acrylamide	0.070	0.070	-
agarose	-	-	0.170
chemicals	0.026	0.026	0.015
disposables	0.252	0.252	0.010
GelBond	0.063	-	-
film	-	0.032	0.022
Total:	**1.273**	**1.458**	**1.077**

Table 3: Cost estimates of experimental material in MAAP procedures. Cost estimates are based on current US catalog prices and include materials required by published protocols as performed in our laboratory. Estimates assume 100 % success and do not include costs for labor, instrument wear, communication and DNA extraction.

About two-thirds of the expenses in a DAF reaction come from the costs for the DNA polymerase. This figure excludes the time component of DNA isolation as well as running costs. Several sources of *Taq* DNA polymerase exist. Some of these are home-made preparations, which suffice for large scale screening. However, quality and DNA contamination may vary, resulting in variable results. Once an acceptable result is obtained, it is repeated with the more costly, commercial enzyme.

Table 3 compares the costs of running a DNA profiling reaction for the three single primer methods. The costs are very similar (ranging from US$1.08 to $1.46). However, RAPD and DAF required only half the time (about 6 hours) compared to AP-PCR.

Significant other costs accrue during the preparation of a final report on a DNA profiling. There are telecommunication and mailing costs. Photography is required as is time to evaluate the data and to write the report. Equipment costs are significant as well. The average thermocyler costs about US$4,000. Top-of-the line models may run at US$10,000. Gel electrophoresis equipment and pipetting devices add up to another US$4,000 for normal usage. Report preparation requires desk-top publishing and quality photography or scanning. DNA isolation requires centrifugation and quality waterbaths. Additional costs come from the need to run quality control reactions. Samples are run without template DNA or without enzyme. Amplifications for customers are always run on duplicate DNA isolations and gel runs. Preliminary results frequently are re-run for pairwise comparisons. Regularly the DNA of standard reference material is tested for reliability. Plant material at times needs to be grown in greenhouses to give young growth, needed for better DNA isolation. In short, while the sample cost may hover around one or two dollars, the real labor inputs and costs for infrastructure are significantly larger. With existing technology, lacking automation and robotics, easily places the cost per sample per primer tested at around US$75. Multiple isolations and parallel processing substantially lowers that cost. Since DAF and AFLP tests produce about 50 to 100 data points per analysis, costs are already approaching a comfortable US$1 per data point. Technology development in the near future as well as larger customer demand is clearly going to lower this cost.

Technology transfer is not always in the hands of scientists and inventors, but relies on the combination of a legal, financial and management team with a vision and desire to see technologies reach the commercial marketplace. Universities require technology transfer offices, which in combination with the academic leadership of the university, provide the environment to develop high risk technologies. Incubator space for fledgling companies may be a means of providing help. State governments should recognize that small companies associated with universities provide additional learning opportunities for students, while helping the employment flexibilities of young companies. The integration of academic, government and private research in Japan should provide some guidelines for such developments.

We hope this transfer of technology becomes a reality, so that the needs for fast and low cost DNA diagnostics in developed and developing countries are adequately covered.

Acknowledgements

We thank Mr. Gerard Weiser (Weiser and Assoc., Philadephia) for legal counseling in the patenting process. Several colleagues, graduate students and postdoctoral staff contributed to many aspects of this work. Specifically we want to thank Brant Bassam, Qunyi Jiang, Rekha Prabhu, Kristal Weaver, Lloyd Callahan, Alexander Kolchinsky, Patrick Williams, Farshid Ghassemi, and Shane Abbitt. Research was supported by the Racheff Endowment, and in part by the Tennessee Agricultural Experiment Station, the International Atomic Energy Agency (Vienna), and the United Soybean Board.

References

Baum, T. J., Gresshoff, P.M., Lewis, S.A. & Dean, R.A. (1994) *MPMI* **7**, 39-47.

Bassam, B.J., Caetano-Anollés, G. & Gresshoff, P.M. (1991) *Analytical Biochemistry* **196**, 80-83.

Caetano-Anollés, G. (1994) *Plant Mol. Biol.* **25**, 1011-1026.

Caetano-Anollés, G. & Gresshoff, P.M. (1994a) *Bio/Technology* **12**, 619-623.

Caetano-Anollés, G. & Gresshoff, P.M. (1994b) *Promega Notes* **45**, 13-18.

Caetano-Anollés, G., Bassam, B.J. & Gresshoff, P.M. (1991a) *Bio/Technology* **9**, 553-557.

Caetano-Anollés, G., Bassam, B.J. & Gresshoff, P.M. (1991b) *Plant Mol. Biology Reporter* **9**, 292-305.

Caetano-Anollés, G., Bassam, B.J. & Gresshoff, P.M. (1992a) In: *Application of RAPD Technology to Plant Breeding*. ed. M. Neff. ASHS, publ. (St. Paul, MN). pp 18-25.

Caetano-Anollés, G., Bassam, B.J. & Gresshoff, P.M. (1992b) *Bio/Technology* **10**, 937.

Caetano-Anollés, G., Bassam, B.J. & Gresshoff, P.M. (1992c) *Mol. Gen. Genet.* **235**, 157-165.

Caetano-Anollés, G., Bassam, B.J. & Gresshoff, P.M. (1993) *Mol. Gen. Genetics* **241**, 57-64.

Caetano-Anollés, G. Williams, P., Callahan, L., Weaver, K. & Gresshoff, P.M. (1995) *Theor. Appl. Genetics* **91**: 228-235.

Callahan, L. M., Caetano-Anollés, G., Bassam, B.J., Weaver, K., MacKenzie, A. & Gresshoff, P.M. (1993) *Golf Course Management*. **61**, 80-86.

Carroll, B.J., McNeil, D.L. & Gresshoff, P.M. (1986) *Plant Science* **47**, 109-114.

Chetverin, A.B. & Kramer, F.R. (1994) *BioTechnology* **12**, 1093-1099.

Collins, G.G. & Symons, R.H. (1993) *Plant Molecular Biology Reporter* **11**, 105-112.

Gresshoff, P.M. (1990) *The New Biologist* **2**, 107-109.

Gresshoff, P.M. (1992) *Grower Talks* (July 92), pp 119-127.

Gresshoff, P.M. (1993) *Plant Breeding Reviews* **11**, 275-318.

Gresshoff, P.M. & MacKenzie, A. (1994) *Chinese J. Botany* **6**, 1-6.

He, S., Ohm, H. and McKenzie, S. (1992) *Theor. Appl. Genet.* **84**, 573-578.

Kolchinsky, A., Funke, R. P. & Gresshoff, P.M. (1993) *Biotechniques* **14**, 400-403.

Kolchinsky, A., Landau-Ellis, D. & Gresshoff, P.M. (1997) *Molec. Gen. Genetics* (in press).

Kruchinina, N.G. and Gresshoff, P.M. (1994) Detergent affects silver sequencing. *Biotechniques* **17**, 279-282.

Landau-Ellis, D., Angermüller, S. A., Shoemaker, R., & Gresshoff, P.M. (1991) *Mol. Gen. Genet.* **228**, 221-226.

Lark, K.G., Weisemann, J.M., Mathews, B.F., Palmer, R., Chase, K. & Macalma, T. (1993) *Theor. Applied Genetics* **86**, 901-906.

Michelmore, R.W., Paran, I. & Kesseli, R.V. (1991) *Proc. Natl. Acad. Sci. (USA)* **88**, 9828-9832.

Mullis, K.B. (1991) *PCR Meth. Applic.* **1**, 1-4.

Paran, I. & Michelmore, R.W. (1993) *Theoretical and Applied Genetics* **85**, 985-993.

Prabhu, R.R. & Gresshoff, P.M. (1994) *Plant Molecular Biology* **26**, 105-116.

Prabhu, R., Jessen, H., Webb, D. Luk, S., Smith, S. and Gresshoff, P.M. (1997) *Crop Science* (in press).

Rafalski, J.A., Hanafey, M.K., Tingey, S.V. & Williams, J.G.K. (1994) In: *Plant Genome Analysis.* ed. P.M. Gresshoff, CRC Press, Boca Raton, Fl., pp 19-29.

Shoemaker, R.C. & Olson, T.C. (1993) In: *Genetics Maps; Locus Maps of Complex Genomes.* ed. S.J. O'Brien. Cold Spring Harbor Press, New York. pp 6.131-6.138.

Smith, J.S.C., Zabeau, M. & Wright, S. (1993) *Maize Genetics Newsletter* **67**, 62-64.

Vos, P., Hogers, R., Reijans, M., van de Lee, T., Hornes, M., Frijters, A., Pot, J., Peleman, J., Kuiper, M. & Zabeau, M. (1995) *Nucl. Acids Res.* **23**, 4407-4414.

Weaver, K., Caetano-Anollés, G., Gresshoff, P.M. & Callahan, L.M. (1994) *BioTech.niques* **16**, 226-227.

Weaver, K., Caetano-Anollés, G., Gresshoff, P.M. & Callahan, L.M. (1995) *Crop Science* **35**: 881-885.

Welsh, J. & McClelland, M. (1991) *Nucleic Acids Res.* **18**, 7213-7218.

Williams, J.G.K, Kubelik, A.R, Livak, K.J, Rafalski, J.A. & Tingey, S.V. (1990) *Nucleic Acids Res.* **18**, 6531-6535.

Phylogenetic Relationships in the Tribe Triticeae

Ponaka V. Reddy and Khairy M. Soliman

Department of Plant and Soil Science, Alabama Agricultural and Mechanical University, Normal, AL 35762-1208, USA

Introduction

The utilization of cereals has had a profound influence on the history of our species. The impact of cereals can hardly be overestimated. Since the first agricultural revolution nearly 10,000 years ago, they have been the most important staple food of humans. They are easy to store and transport and have been and are an important commodity in world trade. The most important cereals in the world are wheat, rice, and maize and there are many other cereals of lesser importance. Within the grass tribe Triticeae, wheat, barley, and rye are the most important cereals. In addition to these very well known species, other species in the Triticeae tribe are useful as forages or possess agronomically valuable traits for wheat and barley improvement (e.g., disease-resistance genes, perennial habit, salt and drought tolerance, early ripening, and dwarfing genes).

Species belonging to the genera *Agropyron, Aegilops, Elymus, Elytrigia,* and *Psathyrostachys* of the Triticeae are potential sources of genetic variation that may be useful in the improvement of related species such as *Triticum, Hordeum* and *Secale.* Early efforts in this direction involved classification studies based on morphological characters. These characters, however, could be affected by developmental stage and environmental fluctuations (Wang and Tanksley, 1989). Isozyme studies provided valuable insights into the phylogeny among the genera and species in the Triticeae (Asins and Carbonell, 1986). However, the number of loci that can be examined are limited, and the data obtained can be developmental stage- and tissue-specific. The most accurate and reliable method of studying phylogenies would be probably by DNA sequencing, but this method is very laborious and expensive, and few genes from Triticeae species have been isolated and sequenced to date. The use of DNA markers in plant systematics represents a new approach that is capable of overcoming most of the above limitations (Crawford, 1990).

The present study aims to clarify the phylogenetic relationships in the tribe Triticeae with two different approaches based on chloroplast DNA restriction fragment variation and with consensus tRNA primer amplification products.

Chloroplast DNA variation and phylogenetic relationships

Restriction fragment length polymorphism (RFLP) analysis has two distinct advantages over other approaches: (1) the absence of tissue and developmental stage problems, and (2) a much larger number of loci available to be tested. In this study we investigated the phylogeny of Triticeae species by comparative analysis of restriction fragment patterns of chloroplast DNA among the crop species and their wild relatives, to find out the origin of crop species such as wheat and barley. The family Gramineae is divided into five subfamilies and 43 tribes, only seven of which contain cereal crops, and tribe Triticeae is one among the seven. This study aims to clarify the phylogenetic relationships among genera and species belonging to tribe Triticeae by the analysis of restriction fragment patterns of their cpDNAs.

Procedures for chloroplast DNA RFLP analysis

Plant material and DNA extraction

Ninety accessions belonging to 52 species and 9 genera (*Psathyrostachys, Leymus, Elymus, Agropyron, Elytrigia, Aegilops, Triticum, Secale,* and *Hordeum*) of the tribe Triticeae, representing both wild and cultivated species, were used in this study. Plants were grown from seeds in a greenhouse at 20°C. Five plants were sampled from each accession and total cellular DNA of the pooled sample was prepared as described by Murray and Thompson (1980) and Saghai-Maroof et al (1984). DNA concentration was determined from the absorbance measured at 260 nm with a Beckman spectrophotometer.

DNA restriction and agarose gel electrophoresis

Restriction enzymes were purchased from Bethesda Research Laboratories, and the restrictions carried out according to the manufacturer's recommendations. Each of the 90 DNAs (Table 1) was restricted with restriction endonucleases *Pst*I, *Hin*fI, *Bgl*II, *Sst*I, *Dra*I, *Kpn*I, *Eco*RI, *Rsa*I, *Hind*III and *Hae*III. Restricted DNAs were electrophoresed in 0.7 % agarose gels with a running buffer of 100 mM Tris-acetate, 1 mM EDTA pH (8.1). DNA fragment sizes were estimated using *Hind*III digested bacteriophage lambda DNA and 1 kb DNA ladder (BRL).

Serial Number	Accession Number	Species Name	Origin
1.	CIae 3	*Aegilops tauschii*	Afghanistan
2.	PI 431603	*Aegilops tauschii*	USSR
3.	PI 542172	*Aegilops comosa*	Turkey
4.	PI 170198	*Aegilops squarrosa*	Turkey
5.	PI 483025	*Aegilops squarrosa*	Cyprus
6.	PI 170203	*Aegilops speltoides*	Turkey
7.	PI 393495	*Aegilops speltoides*	Israel
8.	PI 172357	*Aegilops cylindrica*	Turkey
9.	PI 254864	*Aegilops cylindrica*	Spain
10.	PI 177241	*Aegilops biuncialis*	Turkey
11.	PI 374321	*Aegilops biuncialis*	Yugoslavia
12.	PI 220328	*Aegilops triuncialis*	Afghanistan
13.	PI 487219	*Aegilops umbellulata*	Syria
14.	PI 276993	*Aegilops juvenalis*	Japan
15.	PI 542192	*Aegilops juvenalis*	Turkey
16.	PI 330493	*Aegilops ventricosa*	Japan
17.	PI 298887	*Aegilops markgrafii*	Turkey
18.	PI 542201	*Aegilops markgrafii*	Turkey
19.	PI 317393	*Aegilops crassa*	Afghanistan
20.	PI 487286	*Aegilops crassa*	Jordan
21.	PI 318647	*Aegilops longissima*	Israel
22.	PI 554235	*Aegilops longissima*	Israel
23.	PI 483010	*Aegilops peregrina*	Cyprus
24.	PI 487274	*Aegilops peregrina*	Jordan
25.	PI 486281	*Aegilops columnaris*	Turkey
26.	PI 487198	*Aegilops columnaris*	Syria
27.	CItr 8399	*Triticum aestivum*	Argentina
28.	CItr 7583	*Triticum aestivum*	USSR
29.	CItr 3139	*Triticum durum*	Tunisia
30.	CItr 7668	*Triticum durum*	USSR
31.	CItr 4529	*Triticum sphaerococcum*	USSR
32.	CItr 4571	*Triticum compactum*	USSR
33.	PI 114638	*Triticum compactum*	Australia
34.	PI 283888	*Triticum carthlicum*	Iran
35.	CItr 14787	*Triticum dicoccum*	Ethiopia
36.	CItr 7864	*Triticum turgidum*	USA
37.	PI 134955	*Triticum turgidum*	Portugal
38.	CItr 11802	*Triticum timopheevii*	USSR
39.	CItr 13920	*Triticum hybrid*	Sweden
40.	PI 351663	*Triticum hybrid*	USSR
41.	PI 184543	*Triticum turanicum*	Portugal
42.	PI 317492	*Triticum turanicum*	Afghanistan
43.	CItr 14803	*Triticum polonicum*	Ethiopia
44.	PI 272568	*Triticum polonicum*	Hungary
45.	PI 470724	*Triticum boeticum*	Turkey
46.	CItr 17677	*Triticum araraticum*	Turkey
47.	PI 427319	*Triticum araraticum*	Iraq

Table 1: Different species and accessions used in this study

Serial Number	Accession Number	Species Name	Origin
48.	CItr 17895	*Triticum spelta*	USA
49.	PI 295064	*Triticum spelta*	Hungary
50.	PI 113961	*Triticum macha*	USSR
51.	PI 355508	*Triticum macha*	Syria
52.	PI 119423	*Triticum monococcum*	Turkey
53.	PI 355523	*Triticum monococcum*	Germany
54.	PI 256029	*Triticum dicoccoides*	Spain
55.	PI 414718	*Triticum dicoccoides*	Israel
56.	PI 294478	*Triticum ispahanicum*	France
57.	PI 352492	*Triticum ispahanicum*	Iran
58.	PI 349050	Triticum karamyschevii	USSR
59.	PI 428183	*Triticum uraratu*	USSR
60.	PI 428194	*Triticum uraratu*	Turkey
61.	PI 343190	*Psathyrostachys fragilis*	Iran
62.	PI 401393	*Psathyrostachys fragilis*	Iran
63.	PI 499672	*Psathyrostachys juncea*	China
64.	PI 499675	*Psathyrostachys juncea*	China
65.	PI 236809	*Leymus cinereus*	Canada
66.	PI 478831	*Leymus cinereus*	Canada
67.	PI 531811	*Leymus racemosus*	USSR
68.	PI 531812	*Leymus racemosus*	USSR
69.	PI 232156	*Elymus trachycaulus*	Nevada, USA
70.	PI 232166	*Elymus trachycaulus*	Colorado, USA
71.	PI 172364	*Elymus caninus*	Turkey
72.	PI 252044	*Elymus caninus*	Italy
73.	PI 269406	*Hordeum bogdanii*	Afghanistan
74.	PI 314696	*Hordeum bogdanii*	USSR
75.	PI 401386	*Hordeum bulbosum*	USSR
76.	PI 440413	*Hordeum bulbosum*	USSR
77.	PI 531785	*Hordeum muticum*	Argentina
78.	PI 531787	*Hordeum procerum*	Argentina
79.	PI 297870	*Agropyron cristatum*	Romania
80.	PI 314599	*Agropyron cristatum*	USSR
81.	PI 249143	*Agropyron desertorum*	Portugal
82.	PI 283162	*Agropyron desertorum*	England
83.	PI 174011	*Elytrigia repens*	Turkey
84.	PI 221901	*Elytrigia repens*	Afghanistan
85.	CIse 13	*Secale cereale*	USA
86.	CIse 105	*Secale segetale*	Italy
87.	PI 282604	*Hordeum spontaneum*	USSR
88.	PI 382247	*Hordeum spontaneum*	USSR
89.	DL 71	*Hordeum vulgare*	Argentina
90.	ARLO	*Hordeum vulgare*	Argentina

Table 1: (continued) Different species and accessions used in this study (USSR is used as the then-source of the material)

Transfer and hybridization of DNA fragments

DNA fragments in the gels were denatured in 0.5 M NaOH and 1.5 M NaCl, neutralized in 3 M sodium acetate, and transferred to a Nylon hybridization membrane (ICN Biomedicals) according to manufacturer's recommendations. Nylon filters were prehybridized overnight in a hybridization buffer of 5 x SSC, 10 mM phosphate buffer, 5 x Denhardt's solution, 2.5 mM EDTA, 0.4 % SDS, and 10 mg/ml solution of sonicated non-homologous sperm DNA. Cloned DNA fragments were labeled with ^{32}P-dCTP by nick-translation as described by Sambrook et al (1989). Nick-translated probes were separated from unincorporated ^{32}P-dCTP on spun columns, denatured and then added to the hybridization buffer. Hybridizations were carried out at 65°C overnight. Hybridization membranes were washed at room temperature in 2 x SSC, 0.1 % SDS buffer four times for five minutes each time. Later screens were washed at 50°C in 0.1 x SSC, 0.1 % SDS twice for fifteen minutes each, then the membranes were exposed to X-ray film (Kodak XAR-5) at minus 80°C for 6 hours to 1 week using DuPont intensifier screens. Cloned portions of the chloroplast genome were used to probe the Nylon filters. These included plasmid pBR322 clones of the petunia chloroplast genome, namely Ps 6, Pst8, Pst10 and Pst12.

Stripping probe from blots for reprobing

The blots were unwrapped, placed in a sealable container and strip buffer was added (2 mM DNA phosphate buffer, 2 mM EDTA, 2 mM Tris pH 8.0 and 0.1 % SDS). The blots were agitated gently at 65°C for 10 min and the buffer was discarded. Later fresh buffer was added and incubated at 65°C for 1.5 hours. The buffer was discarded again and fresh buffer was added and incubated for 1 hour at 65°C. Blots were wrapped in Saran plastic wrap and autoradiographed for 2 days to verify probe removal.

Data analysis

Each fragment detected by Southern analysis for each probe/enzyme combination tested was treated as a character, and its presence or absence was scored assuming that common restriction fragments among different taxa were indicative of homologies in genomic DNA sequences. The heuristic approach of the computer program "Phylogenetic Analysis Using Parsimony", version 3.1 (Swofford, 1993) was used on a Macintosh computer to generate the phylogram. The genus *Psathyrostachys* was used as an outgroup because it represented a plant group that has been shown to be well separated within the tribe by previous studies and therefore was useful for phylogenetic comparison.

Results and discussion

Restriction fragment patterns of DNAs were obtained from the 90 accessions of Triticeae species, and compared with each other. As an example of the results the patterns of *Rsa*I digested *Triticum* and *Aegilops* samples electrophoresed on 0.7%

agarose gels and hybridized with Pst8 *Petunia* chloroplast DNA clone are presented in Figure 1. The banding patterns of *Ae. longissima* and *Ae. speltoides* are similar to those of *T. aestivum, T. spelta, T. macha,* and *T. sphaerococcum;* all these four species contain ABD genomes.

Polyploid wheats were classified into three groups: Emmer (AA BB), Dinkel (AA BB DD), Timopheevii (AA GG and AA AA GG). The A and D genome have been postulated to have derived from Einkorn (AA) wheat and *Ae. squarrosa* respectively. But there is diversity of opinion as to the donor of the B genome to the breadwheat and the G genome to Timopheevii.

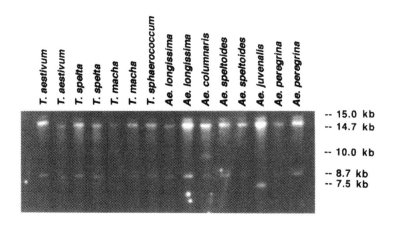

Figure 1: Autoradiographs of Southern blot hybridizations of some of the Triticum *and* Aegilops *species total DNA digested with* RsaI *and hybridized with the nick-translated petunia chloroplast Pst8 clone.*

The origin of the B genome

There are different views about which species donated the B genome to common wheat. Sarkar and Stebbins (1956) showed that *Ae. speltoides* is the B genome donor to breadwheat, based on external morphology, karyotype, and chromosome pairing. Sears (1956) showed that *Ae. bicornis* is the B genome donor to breadwheat based on the morphology of synthetic amphidiploid (SS AA). Johnson (1972, 1975) showed that *T. urartu* is the B donor, based mainly on seed protein profiles. Feldman (1978) showed that *Ae. longissima* is the donor based on karyotype, geographical distribution and chromosome pairing data. The present results showed that among the diploid species of *Aegilops* section sitopsis and *T. urartu, Ae. speltoides* and *Ae. longissima* are close to common wheat as judged by chloroplast DNA restriction fragment patterns.

The origin of the G genome

Based on comparative studies of eu- and alloplasmic Emmer and common wheats, Maan and Lucken (1971) suggested that *Ae. speltoides* is the cytoplasm donor to Timopheevii. Mukai et al (1978) suggested *Ae. aucheri* is the G genome donor. The present results revealed that among other *Triticum* and *Aegilops* species tested, only *Ae. speltoides* is similar to *T. timopheevii* cp DNA. Thus *Ae. speltoides* is the probable donor of the Timopheevii cytoplasm. Figure 2 shows the similar banding patterns of the *T. timopheevii* and *Ae. speltoides* when the DNA samples digested with *Eco*RI and probed with Pst6 clone.

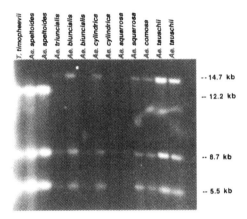

Figure 2: Autoradiographs of Southern blot hybridizations of T. timopheevii *and other* Aegilops *species total DNA digested with* EcoRI *and hybridized with the nick-translated petunia chloroplast Pst6 clone.*

The phylogram (Figure 5) places *H. vulgare* and *H. spontaneum* close to each other. This supports the well accepted hypothesis that the latter is the wild progenitor of the former. The phylogram also shows *H. bogdanii*, which is a central Asiatic species found far from the other species of *Hordeum*. This supports earlier studies concerning the distinctiveness of Asiatic members of the genus based on crossing data, genomic relationships and karyological data. The distinctiveness of *H. bogdanii* and the closeness of *H. vulgare* and *H. spontaneum* can be seen in Figure 3 which shows the DNA samples digested with the restriction enzyme *Kpn*I and hybridized with the petunia Pst12 clone.

The phylogram (Figure 5) shows close associations between *Elymus*, *Elytrigia*, *Leymus*, *Agropyron*, and *Psathyrostachys*. Several crosses have been made between *Psathyrostachys* and *Leymus* species. This intergeneric fertility agrees with the phylogram associations. The similar banding patterns between these species can be seen in Figure 4 which shows the DNA samples digested with *Dra*I and hybridized with the petunia Pst12 clone.

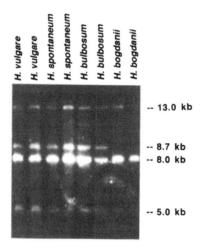

Figure 3: Autoradiographs of Southern blot hybridizations of some of the Hordeum *species total DNA digested with* KpnI *and hybridized with the nick-translated petunia chloroplast Pst12 clone.*

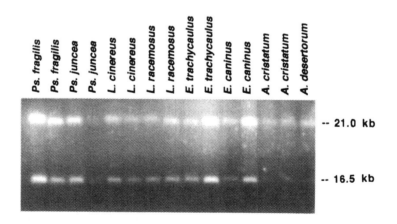

Figure 4: Autoradiographs of Southern blot hybridizations of some of the forage grasses total DNA digested with DraI *and hybridized with the nick-translated petunia chloroplast Pst6 clone.*

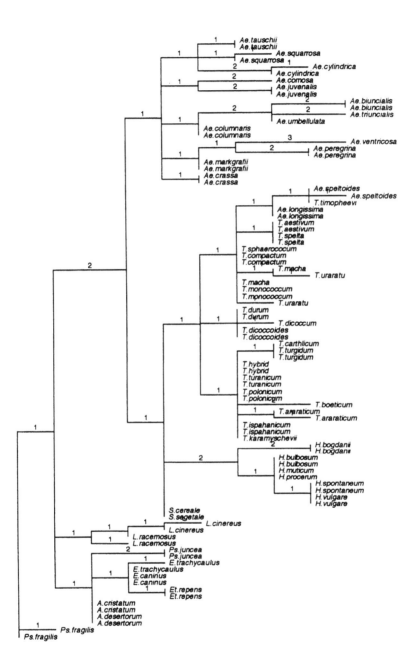

Figure 5: Phylogenetic tree based on chloroplastic markers generated by the PAUP program. Numbers on the branches are the length of branches.

Phylogenetic relationships revealed by PCR using consensus tRNA gene primers

The primary disadvantages of using RFLPs in plant genetic resource characterization are the cost of materials and labor, the processing time and the use of radioactive materials. An alternative method, with the potential to overcome some of the current limitations in determining the genetic relationships is based on the Polymerase Chain Reaction (PCR; Saiki, 1989; Mullis and Faloona, 1987). In this case, polymorphisms are sought in the distance between two short target sequences rather than the presence or absence of restriction endonuclease sites as is the case for standard RFLPs (Skolnick and Wallace, 1988). Oligonucleotides that anneal to the target sequences are used to prime the polymerase reactions. DNA polymorphisms can be detected after amplification using tRNA consensus primers (Welsh and McClelland, 1991). tRNA genes occur in multiple copies and they are dispersed throughout the genome in most species. The DNA encoding tRNAs is the target for amplification, and the amplification products vary as the tRNA gene clusters have evolved (Welsh and McClelland, 1991; Welsh et al, 1991). Here we describe the use of PCR to fingerprint and to show the relationships in the members of tribe Triticeae by using consensus tRNA primers.

Procedures for consensus tRNA PCR analysis

Plant materials and DNA extraction

Forty-one accessions belonging to ten genera (*psathyrostachys, leymus, elymus, agropyron, elytrigia, aegilops, triticum, secale, hordeum* and *triticale*) of the tribe Triticeae were used in this study. DNA extraction was done as described earlier in this chapter.

PCR primer synthesis

Consensus tRNA primers T5B (5' AATGCTCTACCAACTGAACT 3') and T3A (5' GGGGGTTCGAATTCCCGCCGGCCCCA 3') were synthesized by using a Pharmacia Gene Assembler Plus. These primers were developed by Welsh and McClelland (1991) using tRNA sequences, of which over 500 are known.

DNA amplification

Each of the two tRNA consensus primers was used singly for DNA amplification following the protocol reported by Williams et al (1990) with minor modifications. The reaction components were 1x reaction buffer [50 mM KCl, 10 mM Tris-HCl (pH 9.0), and 0.01 % Triton X-100], 200 μM of dNTP, 0.2 μM of primer, 2.5 units of *Taq* polymerase, 50 ng of genomic DNA and 1.0 mM $MgCl_2$ in a 100 μl reaction mix. The *Taq* polymerase, dNTPs and reaction buffer were purchased from Promega, Inc. For the DNA amplification, a Perkin Elmer Cetus

DNA thermal cycler was programmed for 40 cycles at 94°C for 30 sec to denature, 50°C for 30 sec for annealing of primer, and 72°C for 2 min for extension. The reaction mixture was overlaid with mineral oil.

After all the cycles were completed, 30 μl of the samples were loaded in 2.0 % agarose gels in 1x TBE buffer and run at 4 V/cm. A 100 bp DNA ladder (BRL) was used as a molecular standard. The gels were stained in 10 ppm of ethidium bromide solution for 20 min, destained with tap water for 20 min and photographed under UV light with Polaroid type 55 film. The pictures were scanned for amplification products with a gel scanner. Each band was identified by its size in base pairs following the primer used in the reaction. In order to confirm accession-specific markers, each amplification was repeated at least twice. Only the reproducible bands in multiple runs, regardless of their intensity, were considered in the study.

Data analysis

Bands on gels were scored as present "+" or absent "-" for all accessions studied. The sizes of the bands were determined both visually and using a Pharmacia laser densitometer equipped with gel scanning (GSXL) software. The microcomputer package PAUP (Phylogenetic Analysis Using Parsimony), version 3.1, developed by Swofford (Illinois Natural History Survey, Champaign, IL 61820) was used to calculate a pairwise difference matrix and to construct a phylogenetic tree.

Results and Discussion

Identification of markers

The sizes of amplified DNA fragments ranged from 500 to 2,450 base pairs. The number of bands in the profiles varied, depending on the primer and accession tested. No amplification was observed when the primer T3A was used. With the primer T5B the fragments ranged from 2 (X *Triticosecale* sp.) to 12 (*Psathyrostachys fragilis*). Amplification products of 17 accessions with T5B primer is shown in Figure 6. A total of 35 products that could be used as potential genetic markers were disclosed by the T5B tRNA primer. The profiles of the amplified products were compared for identification of species-specific and genus-specific markers (Table 2).

Accession relationships

All pairwise distances among accessions were calculated with the PAUP program. In spite of a wide range in average differences among genera, a smaller trend was observed within the genus. That is, the biggest differences were always found between two genera rather than within the genus. The average marker difference between different genera is presented in Table 3. Thus, the tRNA primer clearly clustered all the species belonging to different genera of the tribe Triticeae.

The table below surveys tRNA markers across 41 accessions. The columns are indexed by S. Number (35 down to 1) with corresponding tRNA Marker sizes; "+" or "−" indicates presence or absence of a band.

S. Number	tRNA Marker
35	500
34	550
33	600
32	625
31	650
30	700
29	750
28	775
27	800
26	830
25	850
24	900
23	925
22	950
21	1000
20	1120
19	1180
18	1230
17	1260
16	1280
15	1450
14	1500
13	1550
12	1600
11	1700
10	1730
9	1780
8	1830
7	1860
6	1990
5	2050
4	2200
3	2250
2	2320
1	2450

Accessions (rows):

1. Ps. fragilis
2. Ps. fragilis
3. Ps. juncea
4. Ps. juncea
5. L. cinereus
6. L. cinereus
7. E. trachycaulus
8. E. trachycaulus
9. E. caninus
10. E. caninus
11. A. cristatum
12. A. desertorum
13. A. fragile
14. A. mongolicum
15. Et. elongitiform
16. Et. pungens
17. Et. pycnantha
18. H. bulbosum
19. H. californicum
20. H. muticum
21. H. procerum
22. H. spontaneum
23. H. spontaneum
24. H. vulgare
25. H. vulgare
26. Ae. tauschii
27. Ae. sp
28. Ae. speltoides
29. Ae. biuncialis
30. T. aestivum
31. T. durum
32. T. spelta
33. T. monococcum
34. T. vavilovii
35. S. cereale
36. S. segetale
37. S. ancestrale
38. S. montanum
39. S. vavilovii
40. X triticosecale
41. X triticosecale

Table 2: Survey of tRNA markers in 41 accessions.
"+" or "-" indicates presence or absence of a band.

The experiments conducted by Welsh and McClelland (1991) indicated that consensus tRNA gene primers that amplify the region between the tRNA genes can be used to generate PCR fingerprints that are generally invariant among strains of the same species and are often substantially conserved among related species. This property made the method applicable to the identification of species by a genome-based method that is independent of other criteria such as morphology. Because this method can be performed easily, and it is independent of genome size, it may be useful for rapidly examining large numbers of different species.

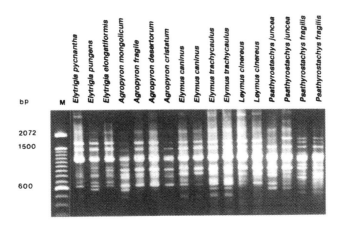

Figure 6: Genomic DNA from 17 Triticeae accessions amplified with consensus tRNA primer T5B.

Most of the polymorphisms disclosed by the primer were useful in the differentiation of the tribe Triticeae into different genera and species. As expected, more monomorphic bands than polymorphic bands were observed within the genus. For example, marker 1280 was present only in the four species of genus *Aegilops*. This suggests a high degree of evolutionary relationships among these accessions. Among the two primers used, one showed 35 markers within the 41 accessions used.

The phylogenetic tree obtained using parsimony separated most of the accessions into their corresponding species and genera (Fig. 7). *Psathyrostachys* was used as an out-group because that genus is considered relatively primitive in the tribe (Baum et al, 1987). The phylogram based on tRNA primers shows a close association among *Psathyrostachys*, *Leymus* and *Elymus* species. Several crosses have been successfully made between *Psathyrostachys* and *Leymus* species (Dewey, 1984), which is consistent with the associations shown here. The phylogenetic tree also clustered *Agropyron* and *Elytrigia* species together, suggesting a close relationship between these two genera. The phylogram showed a close association among the three genera *Secale*, *Triticum*, and *Hordeum*. The *Triticum-Secale* relationship was closer than the *Triticum-Hordeum* and *Secale-Hordeum* relationship. This supports the fact that *Triticum* and *Secale* are placed in the same subtribe, Triticinae, whereas *Hordeum* is placed in another subtribe Hordeinae.

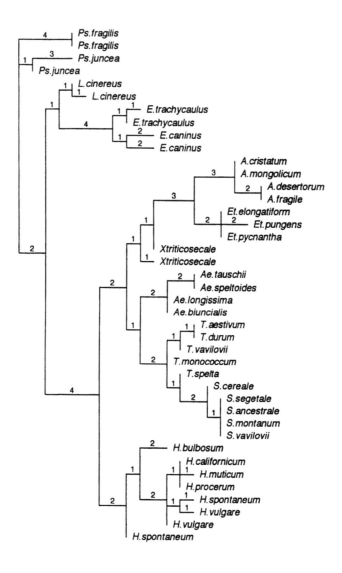

Figure 7: Phylogenetic tree based on consensus tRNA primer amplification products generated by the PAUP program. Numbers on the branches are the length of branches.

The associations between *Triticum* and *Aegilops* were also intense in the phylogram. This agrees with Ogihara and Tsunewaki's (1988) chloroplast DNA studies. Variation within the *Triticum/Aegilops* group was generally found to be low in their study. Thirteen restriction enzymes were needed to detect enough variation to distinguish the various genomic groups. At the generic level, Bowman et al (1983) found no strict demarcation in chloroplast DNA banding patterns between *Aegilops* and *Triticum* and thus they supported the merger of the two genera as proposed earlier by Bowden (1959).

S.No	Name of the Genus	1	2	3	4	5	6	7	8	9	10
1	*Psathyrostachys*	4.2									
2	*Leymus*	7.5	1								
3	*Elymus*	9.8	6.8	3.5							
4	*Agropyron*	12.3	11.5	10.8	1.3						
5	*Elytrigia*	12.9	10.5	12.3	6.3	2					
6	*Hordeum*	12.7	10.5	13	13.9	13.5	3.1				
7	*Aegilops*	12.3	11.5	12.3	12	11.7	8.6	1.3			
8	*Triticum*	12.5	11.3	13.1	11.8	11.5	9.8	5.8	1.6		
9	*Secale*	15.1	10.3	12.1	12.8	12.5	11.2	8.8	4.6	0.4	
10	*Triticale*	10.3	7.5	9.3	7	5.7	8.6	6	11.6	6.8	2

Table 3: Pairwise fragment differences among ten
genera belonging to the tribe Triticeae.

Conclusions

Chloroplast DNA studies, based on RFLP technology, were found to be valuable for comparisons between and within genera, but they were laborious and expensive. In contrast, the PCR-based tRNA single primer amplification was efficient and low cost. Primer T5B provided enough polymorphism to construct a phylogenetic tree between species. Single tRNA primers can be used to compare genomes of organisms at the species/genus level. The consensus results from chloroplast DNA RFLPs, and the tRNA primer PCR revealed that the B genome for bread wheat probably originated from either *Ae. longissima* or *Ae. speltoides*. The G genome of Timopheevii probably originated from *Ae. speltoides*. tRNA primers revealed that *Agropyron* and *Elytrigia* are more closely related to the *Triticum-Aegilops* cluster compared to other forage grasses.

The major clustering of species belonging to *Triticum, Aegilops, Hordeum,* and forage grasses is very similar with both chloroplast DNA and tRNA consensus primer PCR. But there are few differences in the clustering between the two phylogenetic trees as expected. With the tRNA primers amplification products were obtained for the total DNA, which includes chloroplast, mitochondrial, and nuclear DNA. Mitochondrial DNA undergoes sequence rearrangements and nuclear DNA is subject to recombination. Because of these mechanisms, these two DNA classes may not reveal accurate systematic relationships. On the other hand because of its conservative nature, structural stability, and maternal inheritance the systematic relationships obtained with chloroplast DNA are more reliable.

It is unclear whether the tRNA primers worked specifically on tRNA amplicons, or whether they amplified DNA in a DAF-like fashion. Cloning and DNA sequencing of amplification products is required to confirm the genomic relationships.

The here-described phylogenetic studies permit technology transfer, as breeders can evaluate phylogenetic relationships as a guide to introduce 'foreign' genetic material through wide crosses. Recognition of such relationships provides an alternative strategy to gene transfer technology, discussed extensively in other chapters in this volume. Interestingly, gene introduction into a plant through sexual means is as old as agriculture; however, modern molecular techniques provide insights into genomic relationships, thereby giving the plant breeder a proactive, rather than reactive tool to better agricultural productivity.

Acknowledgements

Contributed by the Agricultural Experiment Station, Alabama Agricultural and Mechanical University. The authors wish to thank Dr. David Wagner and Dr. Allan Zipf for their helpful suggestions. This research was supported in part by the National Science Foundation RIMI grant 88-050-97.

References

Asins, M.J. & Carbonell, E.A. (1986) *Theor. Appl. Genet.* **72**, 551-558.

Baum, B.R., Estess, J.R. & Gupta, P.K. (1987) *Amer. J. Bot.* **74**, 1338-1395.

Bowden, W.M. (1959) *Can. J. Bot.* **37**, 657-684.

Bowman, C.M., Bonnard, G. & Dyer, T.A. (1983) *Theor. Appl. Genet.* **65**, 247-262.

Crawford, D.J. (1990) *Plant Molecular Systematics: Macromolecular Approaches.* Wiley-Interscience, New York. pp 13-24.

Dewey, D.R. (1984) *The Genomic System of Classification as a Guide to Intergeneric Hybridization with the Perennial Triticeae.* Gustafson, J.P. (ed.). Plenum Publ. Corp., New York. pp 209-280.

Feldman, M. (1978) *Proc. V^{th} Int. Wheat Genet. Symp.* **1**, 120-132.

Johnson, B.L. (1972) *Proc. Natl. Acad. Sci.* USA **69**, 1398-1402.

Johnson, B.L. (1975) *Can. J. Genet. Cytol.* **17**, 21-39.

Maan, S.S. & Lucken, K. (1971) *J. Hered.* **62**, 149-152.

Mukai, Y., Maan, S.S., Panayotov, I. & Tsunewaki, K. (1978) *Proc. Vth Int. Wheat Genet. Symp.* **1**, 282-292.

Mullis, K.B. & Faloona, F. (1987) *Methods in Enzymology* **155**, 335-350.

Murray, M.G. & Thompson, W.F. (1980) *Nucleic Acids Res.* **8**, 4321-4325.

Ogihara, Y. & Tsunewaki, K. (1988) *Theor. Appl. Genet.* **76**, 321-332.

Saghai-Maroof, M.A., Soliman, K.M., Jorgensen, R.A. & Allard, R.W. (1984) *Proc. Natl. Acad. Sci. USA.* **81**, 8014-8018.

Saiki, R.K. (1989) *Principles and Applications for DNA Amplification.* Stockton Press, New York. pp 7-15.

Sambrook, J., Fritsch E.J. & Maniatis, T. (1989) *Molecular Cloning: a Laboratory Manual.* Cold Spring Harbor Laboratory, Cold Spring Harbor, New York.

Sarkar, P. & Stebbins, G.L. (1956) *Am. J. Bot.* **43**, 297-304.

Sears, E. R. (1956) *Wheat Inform. Serv.* **4**, 8-10.

Skolnick, M.H. & Wallace, R.B. (1988) *Genomics* **2**, 273- 279.

Swofford, D.L. (1993) *PAUP: Phylogenetic Analysis Using Parsimony.* Version 3.1 Computer program distributed by the Illinois Natural History Survey, Champaign, Illinois.

Wang, Z.Y. & Tanksley, S.D. (1989) *Genome* **32**, 1113-1118.

Welsh, J. & McClelland, M. (1991) *Nucleic Acids Res.* **19**, 861-866.

Welsh, J., Honeycutt, R.J., McClelland, M. & Sorbal, B.W.S. (1991) *Theor. Appl. Genet.* **82**, 473-480.

Williams, J.G.K, Kubelik, A.R, Livak, K.J, Rafalski, J.A. & Tingey, S.V. (1990) *Nucleic Acids Res.* **18**, 6531-6535.

Isolation of Plant Peptide Transporter Genes from *Arabidopsis* by Yeast Complementation

Henry-York Steiner[1], Wei Song[1,2], Larry Zhang[3], Fred Naider[3], Jeffrey Becker[1] and Gary Stacey[1,2]

[1]*Department of Microbiology and* [2]*Center for Legume Research, University of Tennessee, Knoxville, TN 37996-0845, USA;* [3]*Department of Chemistry, College of Staten Island, City University of New York, Staten Island, New York 10301, USA*

Introduction

Peptide transport systems mediate the uptake of small peptides across the plasma membrane into the cell cytoplasm. These systems are present in bacteria, fungi, plants, and animals (for reviews see Becker and Naider, 1980; Higgins and Payne, 1980; Matthews and Payne; 1980, Payne; 1980, Naider and Becker, 1987; Matthews, 1987; Payne and Smith, 1994). Transport of peptides is accomplished by a specific biochemical process in which peptides (≤ 6 amino acids) are transported by energy-dependent, saturable carriers. The function of peptide transport is primarily for utilization of peptides as sources of nitrogen and amino acids (Becker and Naider, 1980; Matthews and Payne, 1980; Payne, 1980; Adibi, 1987). However, a few reports have shown peptide transport to be involved in bacterial sporulation (Mathiopoulos et al, 1991; Perego et al, 1991; Koide and Hoch, 1994), chemotaxis in bacteria (Manson et al, 1986), and recycling of bacterial cell wall peptides (Goodell and Higgins, 1987).

In plants, peptide transport has been demonstrated in isolated scutella from germinating grains of barley, wheat, rice, and maize and is thought to be involved in supplying amino acids, stored in the form of peptides, from the endosperm to the germinating embryo (Higgins and Payne, 1978a; Sopanen et al, 1978; Salmenkallio and Sopanen, 1989). Peptide transport in germinating barley seeds was demonstrated with non-hydrolyzable, non-physiological peptide substrates, which were accumulated intact and against a concentration gradient (Burston et al, 1977; Higgins and Payne, 1977a,b; Sopanen et al, 1977). This transport system exhibited saturation kinetics and was inhibited by a range of metabolic inhibitors (Higgins and Payne, 1977b). A number of peptides were also

0-8493-8265-3/97/$0.00+$.50

shown to be transported intact into the barley embryo and subsequently hydrolyzed intracellularly (Higgins and Payne, 1978a,c). The plant peptide transport system can transport both dipeptides and tripeptides (Sopanen et al, 1977; Higgins and Payne, 1978b). Two proteins, approximately 66 and 41 kDa, were initially identified by thiolaffinity-labeling as components of the plant peptide transport system in barley grains (Payne and Walker-Smith, 1987). Subsequently, an additional or similar protein of 54 kDa was identified by photoaffinity labeling (Hardy and Payne, 1991).

A number of peptides and peptide-like compounds have been isolated in plant extracts and in the phloem and xylem (Higgins and Payne, 1982). In addition, several toxins produced by phytopathogenic organisms are peptides or peptide-like compounds (Walton, 1990). Though it has not been shown conclusively, it is conceivable that these molecules may require a means of transport for distribution or entry in the plant which could be accomplished by a peptide transporter. In bacterial and yeast systems, peptide transport has also been investigated as a means to transport normally impermeant, toxic compounds coupled to a physiological peptide substrate into a cell (Payne and Smith, 1994; Becker and Naider, 1994). This idea of illicit transport of toxic peptides to inhibit growth of cells could very well be extended to plants in the form of peptide-based herbicides or growth regulators.

In this paper we report the characterization of two plant peptide transport genes, AtPTR2-A and AtPTR2-B isolated by expressing an *Arabidopsis* cDNA library in the yeast peptide transport mutant *ptr2*.

Results

Complementation and sequence analysis of plant peptide transport genes

An *Arabidopsis* cDNA library (Minet et al, 1992) was transformed into the yeast peptide transport mutants PB1X-9B (*Mata ura3-52 leu2-3,112 lys1-1 his4-38 ptr2-2*) and PB1X-2AΔ (*Mata ura3-52 leu2-3,112 lys1-1 his4-38 PTR2::LEU2*) (Perry et al, 1994) and selected on minimal medium containing the dipeptides His-Leu and Lys-Leu. Several clones were recovered, from which two plasmids were isolated, reintroduced into the yeast mutants and shown to restore the peptide transport phenotype (Steiner et al, 1994). The two plasmid inserts were sequenced and were found to encode polypeptides of 610 (AtPTR2-A, originally called AtPTR2 in Steiner et al, 1994) and 585 (AtPTR2-B; Song et al, 1996) amino acids (Figure 1). AtPTR2-B was reported previously as a histidine transporting protein and designated NTR1 (Frommer et al, 1994). Hydropathy analysis indicated that both proteins were hydrophobic with 12 potential transmembrane domains. A search of protein sequence databases using the NCBI BLAST algorithm (Altschul et al, 1990) showed extensive homology to a number of other recently described peptide transporters and a nitrate transporter from *Arabidopsis* (Steiner et al, 1995;

Song et al, 1996). When these sequences were compared using a multiple sequence alignment algorithm (Higgins et al, 1992), these proteins were shown to have a high degree of homology and comprise a new group of transporters called the PTR (*Peptide TRansport*) family (Steiner et al, 1995).

```
AtPtr1  M G S I E E E A R P - - - - - L I E E G L - - - - - - - - - - - - - - - - - - - - - - - - - - - -  17
AtPtr2  M S S I E E Q I T K S D S D F I T S E D Q S Y L S K E K K A D G S A T I N Q A D E Q S S T D E L Q K  50

AtPtr1  - - - - - I L Q E V K L Y A E D G S V D F N G N P P L K E K T G N W K A C P F I L G N E C C E R L A  61
AtPtr2  S M S T G V L V N G D L Y P S P T E E E L A T L F S V C G T I P - W K A F I I I I V - E L C E R F A  98

AtPtr1  Y Y G I A G N L I T Y L - - - - - - - - T T K L H Q C G N V S A A T N V - - - T T W Q G T C Y L T P L  100
AtPtr2  Y Y G L T V P F Q N Y M Q F G P K D A T P G A L N L G E T G A D G L S N F F T F W - - - C Y V T P V  145

AtPtr1  I G A V L A D A Y W G R Y W T I A C F S G I Y F I G M S A L T L S A S V P A L K P A E C I G D F C P  150
AtPtr2  G A A L I A D Q F L G R Y N T I V C S A V I Y F I G I L I L T C T A I P S V I D A G K S M G G F V V  195

AtPtr1  S A T P A Q Y A M F F G G L Y L I A L G T G G I K P C V S S F G A D Q F D D T D S R E R V R K A S F  200
AtPtr2  S - - - - - - - - - - - L I I L G L G T G G I K S N V S P L M A E Q L P K I P P Y V K T K K N G S  233

AtPtr1  - - - - - - - - - - - - F N W F Y F S I N I G A L V S S S L L V W I Q E N R G W G L G F G I P T V  237
AtPtr2  K V I V D P V V T T S R A Y M I F Y W T I N V G S L - - - S V L A T T S L E S T K G F V Y A Y - - -  278

AtPtr1  F M G L A I A S F F F G T P L Y R F Q R P G G S P I T R I S Q V V V A S F R K S S V K V P E D A T L  287
AtPtr2  L L P L C V - - - - F V I P L I I L A V S K T A F T S T L L P P V P S L F - - - - V L V K C S S L L  319

AtPtr1  L Y E T Q D K N S A I A G S R K I E H T D D C Q Y L D K A A V I S E E E S K S G D Y S N S W R L C T  337
AtPtr2  L - - - - - K T N L I - - S K K L N H L - - - - - - - A L L L L E R Y V K - - - - - D Q W - - - -  346

AtPtr1  V T Q V E E L K I L I R M F P I W A S G I T F S A V Y A Q M S T M F V Q Q G R A M N C K I G S F Q L  387
AtPtr2  D L F I D E L K R A L R A C K T F L F Y P I Y W V C Y G Q M T N N L I S Q A G Q M Q T G N V S N D L  396

AtPtr1  P P A A L G T F D T A S V I I W V P L Y D R F I V P L A R K F T G V D K G F T E I Q R M G I G L F V  437
AtPtr2  - - - - F Q A F D S I A L I I F I P I C D N I I Y P L L R K Y - - - N I P F K P L L R I T L G F M F  439

AtPtr1  S V L C M A A A A I V E I I R L H M A N D L G L V E S G A P V P I S V L W - Q I P Q Y F I L G A A E  486
AtPtr2  A T A S M I Y A A V L Q A K I Y Q R G P C Y A N F T D T C V S N D I S V W I Q I P A Y V L I A F S E  489

AtPtr1  V F Y F I G Q L E F F Y D Q S P D A M R S L C S A L A L L T N A L G N Y L S S L I L T L V T Y F T T  536
AtPtr2  I F A S I T G L E F A F T K A P P S M K S I I T A L F L F T N A F G A - - - - - I L S I C I S S T A  534

AtPtr1  R N G Q E G W I S D N L N S G H L - - D Y F F W L - - - - L A G L - - - - - - - - - - - - S L V N  567
AtPtr2  V N P K L T W M Y T G I A V T A F I A G I M F W V C F H H Y D A M E D E Q N Q L E F K R N D A L T K  584

AtPtr1  M A V - - - - - - - Y F F S A - A R Y K Q K K A S S                                                585
AtPtr2  K D V E K E V H D S Y S M A D E S Q Y N L E K A N C                                                610
```

Figure 1: Protein sequence alignment of the plant peptide transporters AtPTR2-A and AtPTR2-B (NB. denoted ATPtr1 and 2 respectively, in figure only). The AtPTR2-A and AtPTR2-B deduced amino acid sequences were aligned with the DNA* program MegAlign (DNA*, Madison, WI) using the Clustal method (Higgins et al, 1992). The sequences were aligned pairwise with the following parameters; window size = 5, diagonals saved = 5, gap penalty = 3, K-tuple = 1, and residue weight table = PAM 250. Only amino acids that are identical are boxed in the alignment. Dashes indicate gaps within the aligned sequences.

Peptide utilization by AtPTR2-A and AtPTR2-B yeast transformants

A number of peptides varying in both composition and length were examined for their ability to support growth of PB1X-9B transformed with AtPTR2-A and AtPTR2-B (Table 1). The yeast transformants grew on all peptides except tri-Lys, and peptides greater than three residues in length. The plant peptide transporters showed a similar pattern of uptake to that of the yeast peptide transporter *PTR2*, except that the plant transporters were able to utilize Met-Met-Leu while the yeast transporter was not (Table 1). The mutant PB1X-9B, as well as PB1X-9B transformed with a null plasmid (YCp50), showed no growth on any of the peptides. The peptide transport-proficient yeast strain PB1X-2A (*Mata ura3-52 leu2-3,112 lys1-1 his4-38 PTR2*), does not utilize Met-Met-Leu, indicating that the uptake of this peptide is due to specificity conferred by the AtPTR2-A and AtPTR2-B genes.

Radiolabeled di-leucine uptake of AtPTR2-A and AtPTR2-B yeast transformants

Both plant peptide transporters transformed into the yeast peptide transport mutant PB1X-9B restored ^{3}H-di-leucine uptake to wild-type levels (Steiner et al, 1994; Song et al, 1996). Uptake of the radiolabeled substrate could be inhibited with 100-fold cold di-leucine, whereas 100-fold excess cold leucine had no effect on the uptake rate of radiolabeled di-leucine, indicating peptide uptake was not via an amino acid transporter. Uptake rates conferred by the plant peptide transporters were consistent with uptake rates of PB1X-9B transformed with the yeast peptide transporter (Perry et al, 1994). Competition with uptake of ^{3}H-di-leucine similar to that observed by Leu-Leu was also seen for Ala-Ala, Ala-Ala-Ala, Ala-Met, Met-Met-Leu, Met-Met, and Leu-Phe indicating a variety of peptides are recognized by these transporters.

	Yeast Strain/Transformants				
Peptides	**PB1X-9B [AtPTR2-B]**	**PB1X-9B [AtPTR2-A]**	**PB1X-9B [PTR2]**	**PB1X-2A (wild type)**	**PB1X-9B [YCp50]**
Ala-Leu	+	+	+	+	-
Ala-Ala-Leu	+	+	+	+	-
Lys-Ala	+	+	+	+	-
Lys-Ala-Ala	+	+	+	+	-
Lys-Lys	+	+	+	+	-
Lys-Lys-Lys	-	-	-	-	-
Leu-Leu	+	+	+	+	-
Leu-Leu-Leu	+	+	+	+	-
Leu-Met	+	+	+	+	-
Met-Met-Leu	+	+	-	-	-
Gly-Leu-Gly-Leu	-	-	-	-	-
Thr-Pro-Arg-Lys	-	-	-	-	-

Table 1. Growth of yeast on peptide substrates. '+' = growth; '-' = no growth.

Toxicity assays on yeast transformants

It had been shown previously that the peptide transport mutant PB1X-9B was resistant to peptides containing the toxic amino acid analog ethionine (Eth) or oxylysine (Oxylys), whereas the peptide transport-proficient strain PB1X-2A was sensitive to dipeptides and tripeptides containing these amino acid analogs

(Island et al, 1991). Transformation of PB1X-9B with the *AtPTR2-A* gene restored sensitivity to the toxic dipeptides Leu-Eth, Lys-Ala-Eth, Oxylys-Gly, and a fourth toxic peptide, leucine-*m*-fluorophenylalanine (Leu-f-Phe) (Table 2). PB1X-9B transformed with AtPTR2-B did not show the same sensitivity to toxic peptides as did AtPTR2-A. The AtPTR2-B transformants showed decreased sensitivity to the toxic peptides (heavy growth within the zone of inhibition), while remaining sensitive to the toxic amino acid analogs ethionine and *m*-fluorophenylalanine.

			Yeast Strains/Transformants			
Peptides	PB1X-9B [AtPTR2-B]	PB1X-2A	PB1X-9B [PTR2]	PB1X-9B [AtPTR2-A]	PB1X-9B [YCp50]	PB1X-9B
f-Phe[2]	22	23	24	21	22	24
Leu-f-Phe	23[3]	31	20	25	0	0
Ethionine	42	39	31	37	41	41
Leu-Eth	25[3]	37	25	24	0	0
Lys-Ala-Eth	30[3]	35	28	30	0	0
Oxalysine	0	0	0	0	0	0
Oxalys-Gly	31[3]	24	25	30	0	0

Table 2: Toxicity assay with toxic peptides and toxic amino acid analogs. Numbers represent zone of inhibition around disk measured in mm. Standard error for data from two independent experiments was ≤5% of the mean. [2]f-Phe=m-fluorophenylalanine. [3]Growth within the zone of inhibition.

It is possible that AtPTR2-B is not expressed as well as AtPTR2-A resulting in the observed difference in sensitivity between the two transformants. However, since both genes are driven off the same constitutive promoter, this seems unlikely. Another possibility is that AtPTR2-B has a slightly different substrate specificity from that of AtPTR2-A such that the toxic peptides are transported at a much lower rate. Another scenario suggests that the toxic peptides are transported into the cytoplasm but are detoxified in some way, possibly not undergoing hydrolysis to release the toxic amino acid analog. The only difference between the genetic backgrounds of the transformants is the introduction of the *Arabidopsis* genes which are under the phosphoglycerate kinase promoter. Therefore, the difference in specificity is unlikely to arise from differences in genetic background although there could be differences in the interaction with other components of the yeast peptide transport system and the products of the *Arabidopsis* peptide transport genes. As of yet, the difference in transport specificity between AtPTR2-A and AtPTR2-B remains unresolved.

Inhibition of *Arabidopsis* seedling root growth by toxic peptides

Arabidopsis plants were capable of transporting peptides, as root growth of 4-day-old seedlings exposed to peptides containing toxic amino acids was inhibited

(Steiner et al, 1994). Root growth inhibited by Ala-Eth or Leu-Eth could not be reversed in the presence of the following compounds: Met, Ala-Ala, Ala-Ala-Ala (Table 3). Ala-Met and Leu-Met successfully competed with and reversed the toxicity of Ala-Eth and Leu-Eth, respectively. Met-Met was able to reverse the toxicity of both Leu-Eth and Ala-Eth though less effectively. Inhibition of root growth was identical with ethionine or peptides containing ethionine (data not shown). Consistent with the toxicity data, both AtPTR2-A and AtPTR2-B are expressed in *Arabidopsis* roots (Steiner et al, 1994; Song et al, 1996) indicating that *Arabidopsis* roots have the potential to transport peptides from the extracellular environment into the plant.

| | Toxic Peptide | |
Competitor	Ala-Eth	Leu-Eth
Met	0	0
Ala-Ala	0	0
Ala-Ala-Ala	0	0
Met-Met	40	50
Ala-Met	70	N.D.
Leu-Met	N.D.	100

Table 3: Percent reversal of toxic peptide inhibition of Arabidopsis *root growth by competitors. Percent reversal is determined by measuring root growth of untreated control vs. roots grown in the presence of toxic peptide and competitor. Competitors were supplied in 10-fold excess over toxic peptide, except Met (20-fold) and Ala-Ala (25-fold). Ala-Eth and Leu-Eth were supplied at 0.19 μM per disk. N.D. = not determined.*

Discussion

The availability of peptide transport mutants of *Saccharomyces cerevisiae* has allowed us to begin to dissect the components of the plant peptide transport system by functional complementation of yeast mutants. This strategy allows rapid identification of plant genes functionally homologous to yeast genes in well-characterized yeast mutants and has recently been used successfully to clone other plant transport genes (e.g., Anderson et al, 1992; Riesmeier et al, 1992; Sentenac et al, 1992; Frommer et al, 1993; Hsu et al, 1993).

Characterization of AtPTR-A and AtPTR2-B reveals some striking similarities to the yeast peptide transporter, PTR2. Both the plant and yeast transport systems lack a strict side-chain specificity. The yeast peptide transport mutant transformed with either *PTR2*, AtPTR2-A or AtPTR2-B is able to take up and utilize peptides with a variety of side chains (Table 1). An exception to the observed broad substrate specificity is the inability of highly basic peptides to be transported such as Lys-Lys-Lys. These results are consistent with previous work on both yeast (Lichliter et al, 1976; Marder et al, 1977) and plants (Higgins and Payne, 1978b; Sopanen et al, 1978). In general, both the yeast and plant peptide transport systems mediate the uptake of peptides which have a hydrophobic character to them (Tables 1 and 3). These peptides contain one or more of the hydrophobic amino acids alanine, leucine, or methionine. This was also shown in earlier work on yeast and the barley system (Higgins and Payne, 1978a; Becker and Naider, 1980) and is consistent with the observation that transport of radiolabeled di-leucine is inhibited by a number of peptides containing hydrophobic residues (Steiner et al, 1994). A striking difference between AtPTR2-B and AtPTR2-A and the yeast peptide transporter is the transport specificity of toxic peptides. Though the mechanism is unknown, AtPTR2-B appears to have a lower transport affinity or capacity for all of the toxic peptides tested (Table 2).

Between the root and the radiolabeled uptake assays, the ability of nontoxic peptides to compete with the ethionine-containing peptides was inconsistent. Ala-Ala and Ala-Ala-Ala did not reverse the toxicity of Ala-Eth or Leu-Eth in root assays, which was unexpected considering that these peptides compete very effectively with radiolabeled di-leucine in the yeast uptake assay. However, this may reflect a difference in expression of AtPTR2-A and AtPTR2-B in their native genetic background vs. that in yeast. There may also be considerable differences in levels of competitor necessary to achieve strong reversal of the toxicity in the seedling assay vs. the radiolabeled uptake assay due to the different time courses of the assays. Moreover, this difference in competitor specificity may indicate that other factors or components are present in the root vs. the yeast, which determine to a greater degree the specificity of which peptides will be transported in plants.

The cloning of two plant peptide transport genes is consistent with previous reports identifying at least two and possibly three proteins associated with peptide transport in barley grains (Payne and Walker-Smith, 1987; Hardy and Payne, 1991). Additional functions of peptide transport in plants, other than a nutritional role as postulated for barley grains, are not known. However, a wide range of peptides and peptide-like compounds are known to occur in plants (Higgins and Payne, 1980, 1982; Steffens, 1990). The effect of toxic peptides on roots of *Arabidopsis* seedlings clearly indicates that peptide transport may have other functions in the plant other than delivery of peptides from the endosperm to the embryo.

Sequence comparison between the plant and yeast peptide transporters shows considerable similarity at the amino acid level (Steiner et al, 1995; Song et al, 1996). Recently, a number of peptide transporters have been isolated and characterized, mainly from eukaryotic organisms, which show significant sequence homology to the yeast and plant peptide transporters (Steiner et al, 1994). The presence of this family of peptide transporters in organisms ranging from bacteria to human attests to the significance of peptide transport as an important biological process retained throughout evolution.

Isolation of the plant peptide transport genes also brings about the possibility of exploiting this system for delivery of toxic or growth promoting substances to plants in a manner analogous to that postulated for human pathogens (Fickel and Gilvarg, 1973; Higgins, 1987; Becker and Naider, 1994). Indeed, a number of plant pathogens secrete toxins in the form of peptides (Walton, 1990). For example, tabtoxin, produced by *Pseudomonas tabaci*, is a modified dipeptide which causes "wildfire" infections of tobacco (Willis et al, 1991; Gross, 1991). Phaseolotoxin, from *Pseudomonas phaseolicola*, which causes halo blight of beans (Mitchell, 1977), is a linear tripeptide and has been shown to be transported by an oligopeptide transporter in bacteria (Staskawicz and Panopoulos, 1980). It is conceivable that modifying the peptide transport system of important crop species could provide the necessary resistance to these toxins and consequently reduce the amount of crop damage caused by their respective pathogens.

References

Adibi, S.A. (1987) *Metabolism* **36**, 1001-1011.

Altschul, S.F., Gish, W., Miller, W., Myers, E.W. & Lipman, D.J. (1990). *J. Mol. Biol.* **215**, 403-410.

Anderson, J.A., Huprikar, S.S., Kochian, L.V., Lucas, W.J. & Gaber, R.F. (1992) *Proc. Natl. Acad. Sci. USA.* **89**, 3736-3740.

Becker, J.M. & Naider, F. (1980) In: *Microorganisms and Nitrogen Sources,* Payne, J.W. (ed.) (New York: John Wiley & Sons, Inc.), pp 257-279.

Becker, J.M. & Naider, F. (1994) In: *Peptide Based Drug Design: Controlling Transport and Metabolism.* Taylor, M. & Amidon, G. (eds.) American Chemical Society Books, Washington, D.C.: American Chemical Society Books.

Burston, D., Marrs, T.C., Sleisenger, M.H., Sopanen, T. & Matthews, D.M. (1977) In: *Peptide Transport and Hydrolysis. A Ciba Symposium*, Elliott, K. & O'Connor, M. (eds.) (Amsterdam: Associated Scientific Publishers), pp 79-98.

Fickel, T.E. & Gilvarg, C. (1973) *Nature New Biol.* **241**, 161-163.

Frommer, W.B., Hummel, S. & Riesmeier, J.W. (1993) *Proc. Natl. Acad. Sci. USA* **90**, 5944-5948.

Frommer, W.B., Hummel, S. & Rentsch, D. (1994) *FEBS Letters* **347**, 185-189.

Goodell, E.W. & Higgins, C.F. (1987) *J. Bacteriol.* **169**, 3861-3865.

Gross, D.C. (1991) *Ann. Rev. Phytopathol.* **29**, 247-278.

Hardy, D.J. & Payne, J.W. (1991) *Planta* **186**, 44-51.

Higgins, C.F. (1987) *Nature* **327**, 655-656.

Higgins, C.F. & Payne, J.W. (1977a) *Planta* **134**, 205-206.

Higgins, C.F. & Payne, J.W. (1977b) *Planta* **136**, 71-76.

Higgins, C.F. & Payne, J.W. (1978a) *Planta* **138**, 211-216.

Higgins, C.F. & Payne, J.W. (1978b) *Planta* **138**, 217-221.

Higgins, C.F. & Payne, J.W. (1978c) *Planta* **142**, 299-305.

Higgins, C.F. & Payne, J.W. (1980) In: *Microorganisms and Nitrogen Sources*, Payne, J.W. (ed.) (New York: John Wiley & Sons, Inc.), pp 211-256.

Higgins, C.F. & Payne, J.W. (1982) In: *Encyclopaedia of Plant Physiology*, N.S., Vol. 14A, Boulter, D. & Parthier, B. (eds.) (New York: Springer, Verlag), pp 438-458.

Higgins, D.G., Bleasby, A.J. & Fuchs, R. (1992) *Cabios* **8**, 189-191.

Hsu, L.-C., Chiou, T.-J., Chen, L. & Bush, D.R. (1993) *Proc. Natl. Acad. Sci. USA* **90**, 7441-7445.

Island, M.D., Perry, J.R., Naider, F. & Becker, J.M. (1991) *Curr. Genetics* **20**, 457-463.

Koide, A. & Hoch, J.A. (1994) *Molecular Microbiology* **13**, 417-426.

Kyte, J. & Doolittle, R.F. (1982) *J. Mol. Biol.* **157**, 105-132.

Lichliter, W.D., Naider, F. & Becker, J.M. (1976) *Antimicrobial Agents Chemotherapy* **10**, 483-490.

Manson, M.D., Blank, V., Brade, G. & Higgins, C. F. (1986) *Nature* **321**, 253-256.

Marder, R., Becker, J.M. & Naider, F. (1977) *J. Bacteriol.* **131**, 906-916.

Mathiopoulos, C., Mueller, J.P., Slack, F.J., Murphy, C.G., Patankar, S., Bukusoglu, G. & Sonenshein, A.L. (1991) *Molec. Microbiol.* **5**, 1903-1913.

Matthews, D.M. (1987) *Contr. Infusion Ther. Clin. Nutr.* **17**, 6-53.

Matthews, D.M. & Payne, J.W. (1980) *Curr. Top. Membr. Transp.* **14**, 331-425.

Minet, M., Dufour, M.-E. & Lacroute, F. (1992) *Plant J.* **2**, 417-422.

Mitchell, R.E. & Bieleski, R.L. (1997) *Plant Physiol.* **60**, 723-729.

Naider, F. & Becker, J.M. (1987) In: *Current Topics in Medical Mycology*, Vol II, McGinnis, M.M. (ed.) Springer Verlag, New York. pp 170-198.

Payne, J.W. (1980) In: *Microorganisms and Nitrogen Sources*, Payne, J.W. (ed.) (New York: John Wiley & Sons, Inc.), pp 211-256.

Payne, J.W. & Walker-Smith, D.J. (1987) *Planta* **170**, 263-271.

Payne, J.W. & Smith, M.W. (1994) *Adv. Micro. Physiol.* **36**, 1-80.

Perego, M., Higgins, C.F., Pearce, S. R., Gallagher, M.P. & Hoch, J.A. (1991) *Molec. Microbiol.* **5**, 173-185.

Perry, J.R., Basrai, M.A., Steiner, H.-Y., Naider, F. & Becker, J.M. (1994) *Molec. Cell Biol.* **14**, 104-115.

Riesmeier, J.W., Willmitzer, L. & Frommer, W.B. (1992) *EMBO J.* **11**, 4705-4713.

Salmenkallio, M. & Sopanen, T. (1989) *Plant Physiology* **89**, 1285-1291.

Sentenac, H., Bonneaud, N., Minet, M., Lacroute, F., Salmon, J.-M., Gaymard, F. & Grignon, C. (1992) *Science* **256**, 663-665.

Song, W., Steiner, H.-Y., Zhang, L., Naider, F., Stacey, G, & Becker, J.M. (1996) *Plant Physiol.* **110**, 171-178.

Sopanen, T., Burston, D. & Matthews, D.M. (1977) *FEBS Letters* **79**, 4-7.

Sopanen, T., Burston, D., Taylor, E. & Matthews, D.M. (1978) *Plant Physiology* **61**, 630-633.

Staskawicz, B. J. & Panopoulos, N.J. (1980) *J. Bacteriol.* **142**, 474-479.

Steffens, J.C. (1990) *Annu. Rev. Plant Physiol. Plant Mol. Biol.* **41**, 553-575.

Steiner, H.-Y., Song, W., Zhang, L., Naider, F., Becker, J.M. & Stacey, G. (1994) *Plant Cell* **6**, 1289-1299.

Steiner, H.-Y., Naider, F. & Becker, J.M. (1995) *Mol. Microbiol.* **16**, 825-834.

Tsay, Y.-F., Schroeder, J.I., Feldmann, K.A. & Crawford, N.M. (1993) *Cell* **72**, 705-713.

Walton, J.D. (1990) In: *Biochemistry of Peptide Antibiotics*, Kleinkauf, H. & von Dohren, H. (eds.) (New York: Walter de Gruyter), pp 179-203.

Willis, D.K., Barta, T.M. & Kinscherf, T.G. (1991) *Experientia* **47**, 765-771.

Confocal Laser Scanning Light Microscopy with Optical Sectioning: Application in Plant Science Research

Sukumar Saha, Anitha Kakani, Val Sapra, Allan Zipf
and Govind C. Sharma

Department of Plant and Soil Science, Alabama A&M University,
P.O. Box 1208, Normal, AL 35762, USA

Introduction

Biological specimens are scarcely ideal for optical examination due to their structural heterogeneity, light absorption properties, refractive index and thickness. Such inherent attributes create serious problems in an investigation using any light microscope. Light and transmission electron microscopy (TEM) have traditionally been used to reveal the morphological changes associated with cellular development. Although in recent years multi-element objective lenses in modern light microscopes have minimized image distortion, nevertheless the limited depth of focus leads to errors in recording two-dimensional images. TEM is a very valuable tool which provides information at the cellular level, but preparation and sectioning of the tissues is a tedious and labor intensive process. Recently, the development of the confocal laser scanning microscope (CLSM) and its digital image acquisition system has made it possible to provide an alternative to other types of microscopy. In confocal microscopy both illumination and detection are confined to a single point in the specimen by inserting spiral filters (usually pin-holes) into the optical paths of the objective and condenser lenses. Confocal imaging almost completely eliminates out-of-focus interference from epifluorescence and produces clean optical sections that can be less than one micron thick. The motorized control focus system and viewing the object on the high resolution monitor also allow scientists to reduce the physical demands of observing samples under a microscope.

The first commercial confocal microscope was available for use in 1987 (Murray, 1992) and recently, it has been used in studies of human, animal (Murray, 1992; Walsh et al, 1991) and plant cells (Schultz, 1992; Fricker et al, 1992). The ability to observe fine details in a thick specimen makes the confocal microscope particularly suitable to study living or fixed preparations. The technical advantage of the CLSM is complemented by the powerful image analysis software accompanying almost all of the commercially available confocal microscopes. For example, the image can be stored on the computer hard disk and later edited at a suitable time.

Research at Alabama A&M University utilizes confocal microscopy to study pollen morphology and further to detect the chromosomal location of genes that affect microgametophyte development in cotton as well as to provide new avenues for evaluating and quantifying fiber quality.

Pollen studies

Research to understand and evaluate pollen structure, physical properties, developmental processes, genetic control and regulatory mechanisms will help scientists to genetically manipulate the processes of cotton microsporogenesis. Several methods have been used in cotton to detect pollen viability based on stain intensity. However, the mechanism by which the non-viable pollen compete with viable pollen in effecting fertilization has not been investigated in detail. This is primarily due to the non-availability of suitable methods to characterize the detailed morphology of pollen grains. Confocal microscopy, with its various imaging modes and 3D-imaging capabilities, is particularly well-suited to detect the spatial organization of pollen by scanning at various pre-specified levels. Besides the obvious value of CLSM for pollen morphology, the sensitivity of this technique has great potential for *in situ* localization of cell components as well as surface location or distribution of stains or fluorescent antibodies, which will have great value in the field of medicine to study pollen allergens.

Procedures

Pollen from inbred line *Gossypium hirsutum* cv. Texas Marker 1 (TM1) and inbred *G. barbadense* cv. Pima 3-79 were used as standards to compare with pollen from monosomic and monotelodisomic F$_1$ plants resulting from interspecific crosses between *G. hirsutum* and *G. barbadense*. The pollen were collected from plants grown under controlled conditions in the greenhouse. Specimens were prepared by collecting flowers at anthesis, dusting pollen on a slide and adding two to three drops of the fluorochrome staining solution. Fluorochrome stains used were fluorescein diacetate (FCR) (Gwyn and Stelly, 1989), acridine orange (Verma and Babu, 1989), rhodamine phalloidin (Fishkind and Wang, 1993) and YOYO-1 (Molecular Probes, Inc., Eugene, OR).

The pollen morphology was studied using a Bio-Rad MRC 600 laser scanning confocal microscope (BIO-RAD Microscience Division, Cambridge, MA). The microscope configuration included a scanning head attached to an upright Olympus BHS epifluorescence microscope allowing rapid switching from conventional to confocal imaging. A Krypton/Argon ion mixed gas laser operating at 480 nm/560 nm/647 nm was used as the excitation source. The images were photographed directly from the computer (Gateway 2000 486/33C) display screen using the Image Corder Plus Camera with EKTAR 100 extra sharp color film.

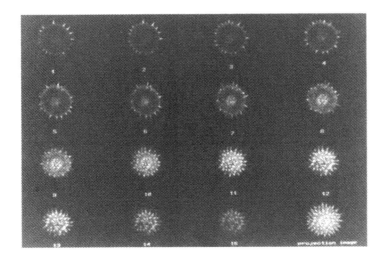

Figure 1: Optical sectioning using the confocal microscope. Fifteen optical sections (5μm thick each) were superimposed to create a composite image of a normal cotton pollen grain.

Results and Discussion

1. Pollen morphology

Figure 1 illustrates the basic design and operation principles of the confocal microscope showing its advantages over conventional light microscopy. The three-dimensional reconstruction of the fluorescently stained pollen grain was assembled by recording a set of images taken at sequential focal planes throughout the grain (optical sectioning). By analogy with physical sectioning techniques, optical sectioning eliminates the contribution of out of focus areas in the specimen image, improving the contrast and visibility of fine details of the specimens. A computer work station generated a 3-D reconstruction of the image superimposing the optical sections and thus providing a detailed view of the ultrastructure of the pollen (Figure 1).

As exemplified, the 3-D images in CLSM provide greater detail and a more accurate spatial relationship of the cellular components. Since all biological entities are essentially three-dimensional, CLSM invariably provides a more realistic rendition of their form and structure.

Cotton pollen is autofluorescent due to the sporopollenin but the fine details of pollen grain structure can be revealed using different fluorochrome stains. Normal pollen grains stained with acridine orange were not distinguishable from those stained with FCR both in stain intensity and stain distribution. However, pollen stained with rhodamine phalloidin seemed to accumulate stain at the base of the spines while the nucleus can clearly been seen in a pollen grain stained with YOYO-1 (Figure 2).

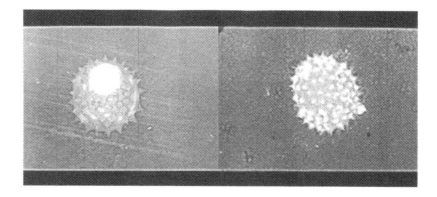

Figure 2: Cotton pollen grain stained with rhodamine phalloidin (right) and YOYO-1 (left). Note the accumulation of stain at the base of the spines in the rhodamine-stained grain and presence of nucleus in the YOYO-1-stained grain.

Our preliminary results indicated that in surface view normal pollen in cotton was spherical in shape with uniform spines and many pores arranged around both poles (Figures 1 and 2).

2. Distinguishing interspecific variation

G. hirsutum (TM1) and *G. barbadense* (Pima 3-79) were readily distinguished based on the pollen morphology. Pollen in TM1 were significantly smaller in size in comparison to Pima 3-79 (Figure 3). The diameter of TM1 pollen averaged 100.9 μm vs. 117.8 μm in Pima 3-79 (Table 1). However, spines on pollen of *G. barbadense* (Pima 3-79) were bigger in size but lesser in number in comparison to

TM1 *G. hirsutum* (Figure 3). The spine size in Pima 3-79 averaged 15.6 mm in comparison to 12.3 mm in TM1. The number of spines were 4.9 per 1,000 mm^2 in Pima 3-79 as opposed to 6.9 per 1,000 mm^2 in TM1 (Table 1).

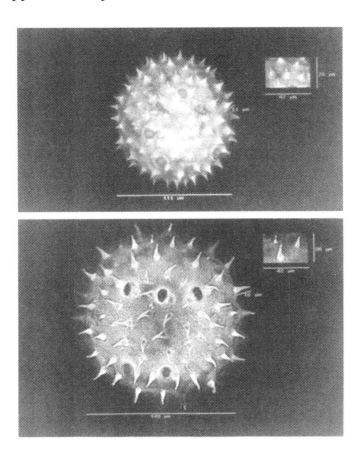

Figure 3: Composite confocal image of the pollen of G. hirsutum, *TM1, (upper) and* G. barbadense, *Pima 3-79, (lower). Note the difference in spine distribution and spine size.*

3. Genetic analysis

Both of the parental inbred lines produced uniform pollen as per their characteristics as discussed above. However, interspecific normal F_1 plants produced three different types of pollen: 1) a smaller TM1 type with dense spines; 2) a larger Pima 3-79 type with dispersed spines; and 3) abnormal pollen without spines (Figure 4).

Distorted segregation ratios in terms of spine distribution and size of the pollen were obtained which are characteristic of interspecific F_2 progenies of (TM1) *G. hirsutum* and (Pima 3-79) and *G. barbadense* (Endrizzi et al, 1984).

	TM-1 (*Gossypium hirsutum*)	PIMA 3-79 (*Gossypium barbadense*)	TM-1 x PIMA 3-79 (F$_1$)		
			Densely distributed spines	Sparsely distributed spines	Without spines
Equatorial Axis (μm) n = 30	100.9 (95 - 114)[1] SD = 4.95	117.8 (85 - 142) SD = 8.29	99.7 (87 - 112) SD = 5.34	107.1 (81 - 119) SD = 8.17	90.8 (77 - 100) SD = 6.69
Average No. of spines/ 10^3 μm^2 n = 75	6.9 (5 - 11) SD = 1.32	4.9 (3 - 9) SD = 1.40	6.5 (5 - 12) SD = 1.70	4.9 (4 - 9) SD = 1.29	NA
Individual spine length (μm) n = 100	12.3 (10 - 15) SD = 1.59	15.6 (12 - 18) SD = 1.63	13.3 (10 - 16) SD = 2.08	13.2 (9 -16) SD = 2.07	NA

Table 1: *Morphological characterization of pollen in cotton according to spine pattern. Range of values,* NA = *not applicable;* SD = *standard deviation.*

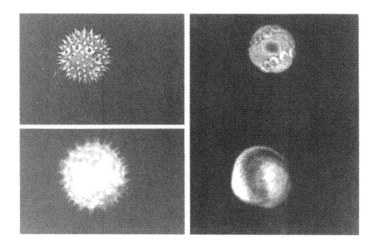

Figure 4: *Pollen types found in F$_1$ plants (TM1 X Pima 3-79). Smaller TM1 type (upper left), larger Pima 3-79 type (lower left) and abnormal pollen without spines (upper and lower right).*

Fluorescence emitted by specimens is the source of optical information used to construct confocal microscopy images. CLSM allowed us to document the brightness of the fluorescence using its image analysis system to aid in the classification of the different types of pollen. Quantitative measurements of fluorescence intensity provide precise image determination of fluorescence marker distribution in three dimensions.

The FCR method revealed variability not only in pollen morphology but also in fluorescence intensity. Viable pollen grains in cotton have been reported to be large, fully engorged with starch and fluoresced brightly compared to non-viable pollen which were small and fluoresced bright to dim to not at all (Gwyn and Stelly, 1989). It is commonly believed that most pollen from normal plants fluoresced brightly whereas interspecific F_1 plants produced more partial to non-fluorescent, non-viable pollen. The image analysis system of the confocal microscope provided the opportunity for determining the intensity of a particular fluorescent stain at an individual pixel level.

Differences in pixel intensity in pollen between the interspecific F_1 plants and normal plants were also observed. However, the abnormal pollen fluoresced more brightly than normal pollen (Figure 5).

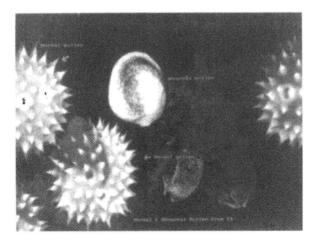

Figure 5: Fluorescein-stained normal and abnormal pollen from F_1 cotton plants. Note the difference in fluorescent intensity.

The apparently high level of stained pollen from heterozygous translocation stocks of cotton (Aslam et al, 1964) also raised the question of the validity of pollen stainability as an indicator of functional pollen (Barrow, 1983).

We are using cytogenetic deficient stocks of cotton to locate genes responsible for pollen development. The effect on the pollen of the genetic loci of the monosomic chromosome can be detected by comparing the pollen morphology of the nullisomic with the normally haploid microspores using CLSM. Compared to normal, spherical pollen, the pollen from a plant deficient for chromosome 12 produced abnormal pollen grains which were semi-spherical in shape with very few spines unevenly distributed about the surface (Figures 6 and 7) suggesting that chromosome 12 in cotton has some very important genes that regulate cotton pollen grain morphology.

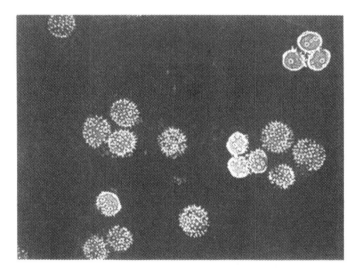

Figure 6: Pollen from H-12$_{F1}$ (chromosome 12 deficient) cotton plants. Note the normal, spiny pollen grains mixed with abnormal pollen.

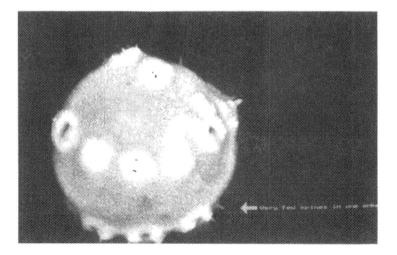

Figure 7: Abnormal pollen type from H-12$_{F1}$ (chromosome 12 deficient) cotton plants. Note the uneven distribution of spines.

Fiber studies

The aggregate value of cotton fiber grown in the USA in 1990 was about US$4 billion. The high value per acre of cotton and the demands for increased uniformity, strength and high quality of fibers clearly justify the need for new and innovative approaches toward evaluating and understanding fiber quality.

The quality of cotton fiber is currently being evaluated, to a large extent, by color, trash content, and various properties of the fibers such as length, strength, elongation, fineness, maturity and micronaire reading. Cotton must be carefully selected and blended to minimize problems associated with lack of fiber uniformity in the yarn and cloth manufacturing process (Dever and Gannway, 1992).

In addition to its commercial importance, cotton fiber is a good experimental system for studying cell elongation and cell wall development because a large population of single cells undergo developmental events in a relatively synchronous manner. The fiber originates and elongates as a single ovular epidermal cell, providing an excellent cell system to study plant cell extension and wall maturation.

We are utilizing the confocal microscope to study the detailed inter- and intra-cellular changes during fiber development of cotton fiber mutants using fluorescent dyes. We believe that the optical sectioning and digital image acquisition system in confocal microscopy will greatly enhance the analysis of fiber morphology and thus contribute to fiber quality. This generates a direct application of the recent advances in laser microscopy to industry.

Procedures

Cotton ovules were collected at different times pre- and post-anthesis, hand sectioned, stained as above and analyzed for fiber development. Mature fibers were collected from mature cotton bolls and stained with the same procedures as for cotton pollen.

Results and discussion

The development of cotton fiber cells typically follows four stages: initiation, elongation, secondary thickening and maturation with the final length of the cotton fiber being the product of the rate of elongation per day and the total period of elongation, which is a genetic attribute (Basra and Malik, 1984). The ability to observe fine details in a thick specimen and the imaging approach to select particular types of cells make the confocal microscope well-suited to study cotton fiber cells. Our studies with ovules and seeds collected from different stages indicated that all of these developmental stages can very easily be visualized with CLSM (Figure 8). CLSM allowed us to detect different types of epidermal cells which confirmed that although all of the epidermal cells were potential fibers not all differentiated into fibers within the first two days of anthesis.

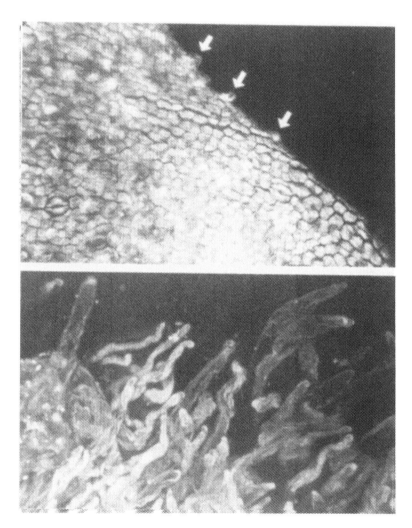

Figure 8: Stages of fiber development from the cotton ovule. Fiber initiation (upper), fiber elongation (lower).

Berlin (1977) documented with electron microscopy that the ratio of fiber to epidermal cells varied tremendously within the cultivars. However, the EM process was very costly, time consuming and laborious. With CLSM several specimens can be screened within a few minutes.

It had been reported that fine fibers with a small diameter and fully developed walls were desirable because they produced the strongest yarn (Basra and Malik, 1984). The digital image analysis system with its numerical operation capabilities allowed us to determine both quantitatively and qualitatively different cytoskeletal parameters including fiber length and diameter of the cell (Figure 9).

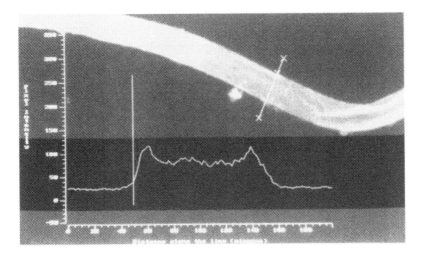

Figure 9: Cytoskeletal parameters of cotton fiber using CLSM. Bar denotes the position along intensity profile.

In addition to cytoskeletal parameters CLSM allowed us to identify the cytoskeletal chemistry of the fiber cells using different stains. A single fiber cell stained with two different stains can be viewed independently or at the same time under two different visual filters, specific for each stain (Figure 10).

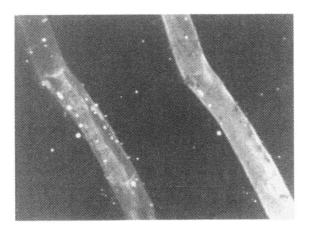

Figure 10: Dual stained cotton fiber cell viewed under rhodamine (left) and fluorescein (right) filter blocks.

The color balance can also be shifted so as to simulate a three-dimensional image when viewed using "3-D glasses". The two images can further be superimposed to create a single image revealing the detailed morphology of the individually stained regions.

Future potential

The sensitivity of confocal microscopy will be exploited most dramatically in the *in situ* localization of individual cell components. In particular, single copy genes can be located to specific chromosomes using fluorescently-labeled oligonucleotides (RNA or DNA) as probes, a technique popularly known as FISH (Fluorescent *in situ* hybridization). CLSM was chosen to localize the knob-specific fluorescent probe on maize meiotic chromosomes (Makowski and Ruzin, 1994). The chromosomes are background stained with one fluorochrome and the probe is labeled with a fluorochrome of a different emission wavelength (color) and the images superimposed to provide the composite picture.

Fluorescently-labeled oligodeoxynucleotides can also be utilized as primers for *in situ* polymerase chain reaction (PCR) amplification and detection. Obviously, this strategy is not restricted to locating the position of single copy genes exclusively. CLSM can pin-point fluorescent antibody and nucleotide probes so that tissue expression as well as cellular compartmentalization can be determined.

Depending upon the microscope system, tissues can be doubly or triply labeled to evaluate the position and distribution of two or three specific cellular components within the same tissue and at the same time. In fact, the ability of CLSM to easily track fluorescence through three dimensions has been underutilized for the study of plant systems, e.g., pollen tube growth, ion accumulation, xylem/phloem source/sink flux, pathogen infection, pest resistance, to name but a few processes. The only limitation has been the development of fluorochromes specific for particular plant cellular components, e.g., waxes (cutin), lipids/fatty acids and polysaccharides other than callose.

Acknowledgements

The authors gratefully acknowledge the financial support of CSRS/USDA (Grant #91-38814-6230) and Cotton Incorporated for this work.

References

Aslam, M., Brown, M.S. & Kohel, R.J. (1964) *Crop Science* **4**, 508-510.

Barrow, J.R. (1983) *Crop Science* **23**, 734-736.

Basra, A.S. & Malik, C.P. (1984) *Int. Rev. Cytol.* **89**, 65-112.

Berlin, J. (1977) In: *Proc. Belt. Cotton Res. Conf.* pp 50-51. Memphis, TN.

Dever, J.K. & Gannway, J.R. (1992) *Crop Science* **32,** 1402-1408.

Endrizzi, J.E., Turcotte, E.L. & Kohel, R.J. (1984) In: *Cotton*, Kohel, R.J. & Lewis, C.F. (eds.). American Society of Agronomy. Madison, WI. pp 82-129.

Fishkind, D.J. & Wang, Y. (1993) *Journal of Cell Biology* **123,** 837-848.

Fricker, M.D., Blatt, M.R. & White, N.S. (1992) *J. Exp. Bot.* **43,** 25-31.

Gwyn, J.J. & Stelly, D.M. (1989) *Crop Science* **29,** 1165-1169.

Makowski, E.R. & Ruzin, S.E. (1994) *Biotechniques* **16,** 256-261.

Murray, J.M. (1992) *J. Neuropathol. Exp. Neurol.* **51,** 475-487.

Schultz, A. (1992) *Protoplasma* **166,** 153-164.

Verma, R.S. & Babu, A. (1989) *Human Chromosomes: Manual of Basic Techniques.* Pergamon Press. New York.

Walsh, M.F., Ding, J.M., Buggy, J. & Terracio, L. (1991) *Anat. Rec.* **23,** 473-481.

Field Testing of Genetically Engineered Crops: Public-Private Institution Comparisons

Patrick A. Stewart and A. Ann Sorensen

The Center for Agriculture in the Environment, P.O. Box 987, DeKalb, IL 60115, USA

Introduction

Since the 1950s, America's intrigue with technology has waned. The average citizen has become more wary of the negative effects of technology. While the American public has had a generally positive orientation towards scientific and technological growth and progress, there is increased support for control over technological development. Events such as the publication of Rachel Carson's *Silent Spring*, stressing the negative effects of pesticides, and the Three Mile Island nuclear reactor accident, have done little to allay the public's perception of risks associated with new technologies and the perceived inability of scientists and academics to deal with the negative consequences.

However, the economic role of technology has become increasingly important for maintaining competitiveness in world markets. Biotechnology is perceived by many as a panacea for the USA economy that appears to be lagging behind the rest of the world. The importance of biotechnology for the future of the American economy has been underscored by the proclamations and actions of economic and political leaders (Gore, 1994).

Innovative public/private partnerships between academia and business have emerged as a driving force in bringing biotechnology rapidly to the marketplace. The success of these partnerships, however, has raised fears that universities are moving away from their mission of serving the public welfare. This chapter will begin to address those fears by analyzing current research trends to see if concern is warranted. We will first consider crops that have been deregulated by the USDA, allowing for their large-scale planting. We will then look at aspirations for and concerns about technology transfer and public/private partnerships. Next, we will look at small scale field tests of genetically engineered crops, and the use of confidential business information by public and private institutions over the past eight years. We will then conclude by considering the implications of our finding for private and public sectors specifically, and agricultural biotechnology in general.

Agricultural biotechnology: Land of milk and tomatoes

More than any other technology, agricultural biotechnology provides examples of the possibilities and risks associated with technological growth. Scientific and technological advances have led to increased milk production and tastier tomatoes (see Chapter 2, this book), and promise reductions in pesticide use. As of March 1996, thirteen genetically modified crops have been deregulated by the USDA with more waiting in the wings. The deregulated products were either already on the market, in the fields for the first time, or awaiting clearance. These new crops demonstrate the cvariety of plant biotechnology products that will be available in the near future.

Calgene FLAVR SAVR Tomato

As the first plant agricultural biotechnology product to hit the consumer market, considerable controversy has sprung up around this crop. The Calgene FLAVR-SAVR™ tomato relies on an inverted anti-sense transcription gene, which controls the ripening of the fruit (see Chapter 2, this volume). This allows for a slowing of the ripening process, so the tomato can be transported after it has had the chance to ripen on the vine. In turn, the company promises a more flavorful tomato that will last on the shelf longer.

However, as the first product to receive clearance by the federal government, it has met with considerable controversy. Much of the public discussion has been inflamed by Jeremy Rifkin of the Foundation on Economic Trends, who has organized a boycott by chefs and other groups. Oddly, the debate has not been so much over the inverted transcription gene, as it has been over the marker gene, which confers resistance to the antibiotic kanamycin. The use of this gene has raised fears that resistance to the antibiotic might be passed on to bacterial pathogens. Calgene Inc. disputes this claim by noting that the gene's products would be destroyed during digestion (Holden, 1994, 512-513).

Herbicide Tolerant Soybean

Monsanto's first genetically engineered plant is a glyphosate-tolerant soybean. Glyphosate is the active ingredient in Monsanto's Roundup herbicide, which is used mainly to control weeds in corn production. Roundup tolerant soybeans allow farmers to use the herbicide on corn and rotate to soybeans the following year, an approach previously problematic due to carry-over effects on the soybean crop. Crop rotations are considered the primary tool in preventing a build-up of insect, weed, and disease problems between growing seasons. Continuous plantings of corn lead quickly to problems with corn rootworm which in turn dictates continual use of soil insecticides. While farmers are required to pay a US$5 fee per bag along with the seed price, and concerns linger over the European Community's willingness to accept genetically engineered crops, over 20% of the United States soybean acreage is expected to be planted with Roundup tolerant soybeans in 1996 (NBIAP, January 1996).

Bt Potato

Monsanto's insect resistant New Leaf™ potato, which made its debut in the summer of 1995, is expected to be widely planted in 1996. The New Leaf potato utilizes *Bacillus thuringiensis* subspecies *tenebrionis* (B.t.t) to confer resistance to the Colorado potato beetle. As one of the first crops implementing insect resistance strategies from B.t. genes incorporated directly into the plant, there have been concerns over targeted insects evolving immunity. As a result, agronomic strategies have been suggested by academic experts and industrial representatives which include monitoring insect populations for resistance, providing refugia for B.t. susceptible insects to conserve their breeding population, using B.t. proteins with high dose expression and novel action modes, and utilizing multiple tactics, including cultural, biological, and chemical factors to avoid resistance problems (NBIAP News Report, December 1995).

Bt Corn

Ciba Seeds and Mycogen Plant Sciences collaborated on corn containing *Bacillus thuringiensis* subspecies *kurstaki* (B.t.k.), effective against the European corn borer, which is responsible for annual economic losses approaching US$1 billion. This product, to be marketed by Ciba Seeds as Maximizer™ hybrid corn with KnockOut™ built-in corn borer control, is expected to reduce both pesticide usage and the time needed to manage applications of pesticides. While only 400,000 acres will be planted in 1996, Maximizer™ corn will be widely available in the United States in 1997, due to marketing and distribution agreements entered into by Ciba Seeds with Growmark, a cooperative marketer, and Mycogen Plant Sciences with Cargill (Biotech Reporter, December 1995, 3).

Virus Resistant Squash

Asgrow Seed Company's virus resistant squash controls for a major cause of crop losses in squash crops and reduces insecticide use to control the insect vectors. By conferring resistance to watermelon mosaic virus and zucchini mosaic virus, Asgrow, a subsidiary of the Upjohn Company, hopes to capture a major portion of the market. While controversy attached to virus resistance and the possibility of new viruses being created (Falk and Bruening, 1994) delayed its final approval, findings of no significant impact (FONSI) to the environment led to its being deregulated by the USDA in 1995 (USDA, 1995).

Technology Transfer and Public/Private Partnerships:

The potential presence of these products on the market so soon after laboratory testing is an indicator of the increasing speed with which products move from laboratory testing to production (Kenney, 1986, 239-240; Kleinman and Kloppenburg, 1988). In contrast to the 13 to 25 years it has taken in the past to bring agricultural products successfully to the marketplace (Rogers, 1983, 156), these products could potentially be commercialized within six years. The decreased time to commercialization is attributable to both the speed and precision of genetic engineering when compared to traditional plant breeding and to the development of partnerships between academia and industry. The swift movement of technology from invention to production is seen as a necessary condition for the continued success of the national and regional economies.

Funding Concerns

One of the chief mechanisms in getting a product from invention to production stage is the development of closer ties between academia and industry. Many authors have encouraged the development of closer ties between university and industrial sectors and analyzed the outcomes of such partnerships (Dorf and Worthington, 1990; Gray et al, 1987; Johnson and Moore, 1990; Marazita, 1991). This is especially true with agricultural biotechnology (Hardy, 1986; Postlewait et al, 1993) where efforts to institutionalize university-industry ties have been documented in Iowa (Shelley et al, 1988; 1990) and Wisconsin (Kleinman and Kloppenburg, 1988; The Biotechnology Project, 1993).

This need for closer ties between universities and industry has been underscored by the funding situation of many universities. Many universities possess aging facilities with obsolete equipment in an era requiring expensive technology in order to produce relevant research and competitive students (OTA, 1992, 415). A greater emphasis on funding applied science at the federal level has meant the proportion of federal support for basic science has fallen. While agricultural biotechnology, which incorporates both basic and applied aspects of science, has experienced growth in funding at the state level, this has only offset the lack of growth in funding at the federal level (MacKenzie et al, 1992).

These shifts in funding and research priorities have compelled universities to find alternative sources for funding research. These sources are most often found in private industry (Etzkowitz and Peters, 1991:135; Lacy et al, 1990:78-79). Linkages with private industry were encouraged by passage of the Technology Transfer Act of 1986 which permits government scientists to form financially lucrative liaisons with the private sector to promote technology transfer.

Ethical Concerns

Ethical issues associated with the increasing ties between industry and university have been raised, especially in biotechnology where close ties between university researchers and corporations are common. Forty six percent of biotechnology firms support biotechnology research at universities while 33 of the 50 states have university-industry centers for the transfer of biotechnology. Because the research institutions of a university influence the attitudes of its researchers (Lacy and Busch, 1989; Etzkowitz, 1992), one can hypothesize that the research agenda of university scientists will reflect the goals of an institution increasingly reliant on private funding. By being so close to the private sector, the university researcher often becomes a part of the production process, through roles ranging from research relationships and consultantships, to sharing patent rights or holding equity in companies, to management roles (Krimsky, 1991, 277). The possible conflicts of interest between the professor's role as impartial scientist and biotechnological entrepreneurer have been considered in depth by Blumenthal et al (1986), Curry and Kenney (1990), Krimsky (1991), and Krimsky et al (1991) and show that while there is still relative academic freedom, industrial collaboration creates the potential for conflicts of interest.

This quandary is accentuated by the institutionalization of formal ties. The development of university research centers, research parks, and other forms of institutional technology transfer, in addition to the more traditional and informal forms of contact such as consultation, serve to provide research and development for industry and funds for academia (Souder and Nassar, 1990:35). However, these ties may also alter the research agenda of an institution devoted to serving the public welfare. Research suggests that this is the case, with research priorities changing to capture industrial funding (Curry and Kenney, 1990:52; Knudson and Pray, 1991). The question remains as to who the university is to serve, the public or private industry, and if it is reliant on the latter for finances, will it be able to take a disinterested political and ethical stance at odds with its funders (Kleinman and Kloppenburg, 1988:83-84).

Field Release by Public and Private Institutions

To determine if increased private industry involvement in university research has had any effect on university and other public sector institutional research agendas, we have analyzed agricultural crop biotechnology field release data from 1987 until the end of 1995. The field release activity of genetically engineered products in the United States has been recorded in two major databases by the U.S. Department of Agriculture's Biologics, Biotechnology and Environmental Protection Division (BBEPD) since field release regulations went into effect in 1987. The first database, which has been in place since the beginning of field release regulation in 1987, is concerned with permit activity, assessing risks intrinsic to the genetically engineered product and the environment in a cautious case-by-case approach. The second database is concerned with

Technology Transfer of Plant Biotechnology

notifications, a faster, less expensive process put in place at the end of March 1993 to facilitate the field testing of plants seen to present no major threats to ecological safety, i.e., have no weedy relatives or likelihood of transmitting genetic information inadvertently[1]. We combined these databases and made comparisons on the basis of total field release permits/notifications, types of crops being experimented on, types of experiments being conducted, and the presence or absence of confidential business information.

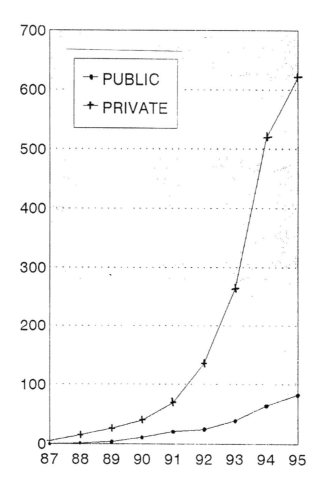

Figure 1: Permits by public and private institutions (APHIS field releases).

---

[1] The crops that were deregulated were corn, cotton, potato, soybean, tobacco, and tomato. The USDA is also proposing to further extend the notification track to include most genetically engineered plants that meet certain eligibility criteria and performance standards (USDA, 1995).

Public and Private Sector Comparisons

It is assumed that the chief difference between the public and private sector is that the public sector is concerned with the extension of knowledge through research and the private sector is concerned with the expansion of profit through innovation. The private sector, which is made up of seventy four chemical, seed, food, start-up biotechnology and other firms, should therefore focus their research on plant modifications that will generate profit, while protecting knowledge gained. On the other hand, the public sector, which is made up of four non-profit or governmental institutions and thirty six universities, should be primarily concerned with extending knowledge and promoting the public trust.

Trends in Field Testing

Overall trends in field testing in the United States show that private industry has taken an increasingly active role in field testing. Since 1987, permit/notification activity has shown exponential growth, although 1995 data suggest that this activity may be tapering off (see Figure 1). The great majority of growth in permits and notifications is by private industry, with public institutions experiencing milder growth patterns. The private sector has accounted for 87 % of all field tests since 1987. The remaining 13 % of the field tests are by the public sector, i.e., universities, non-profit research institutions, and federal and state research centers.

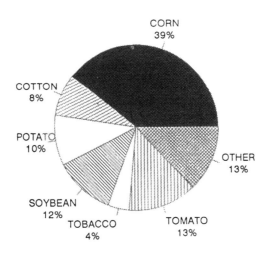

Figure 2: Plant field releases 1987 to 1995.

Types of Crops

Five crops, i.e., corn, cotton, potato, soybean, and tomato, make up 83 % of all field tests conducted between 1987 and 1995 (see Figure 2). These crops are all important production crops in terms of acreage and market share and thus are particularly attractive to commercial ventures.

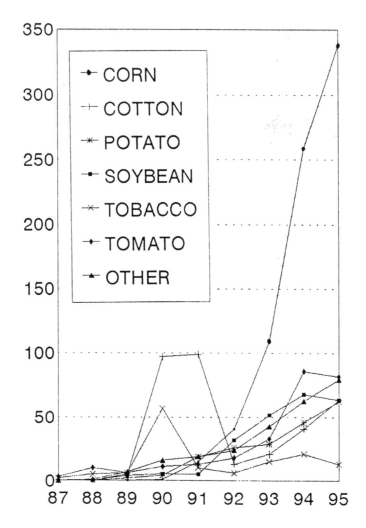

Figure 3: Trends in field testing. Plant permit activity by crop species.

Trends in permits and notifications show that each of the crop categories has tendency toward growth since 1987, with the exception of tobacco, which appears to have leveled off, and soybean and tomato, which have shown a downturn in the last year of data (see Figure 3). Most notable in permit and notification growth has been the case of corn, which has exhibited exponential growth since their first introduction in 1990 when scientific advances with dicots allowed for them to be successfully genetically engineered.

The importance of the major crops is reflected in the emphasis private industry gives to them (see Figure 4). Private industry accounts for all or almost all permit/notification activity in corn, cotton, soybean, and tomato. On the other hand, field tests by public institutions are competitive only in the field testing of potatoes and the broad-based "other" categories.[2]

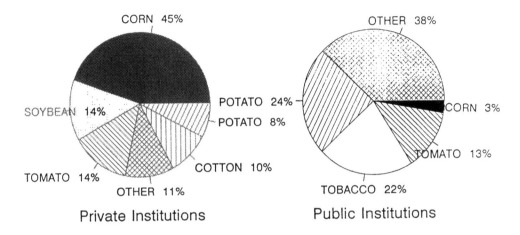

Figure 4: Public and private institutions: Plant field releases 1987 to 1995.

As a percent of total permits and notifications in their discrete categories, the private sector has focused on corn (45 %) and soybean (14 %), two cash crops which tend to be put in rotation. Genetically engineered tomatoes, which have since reached the grocer's shelves and promise to bring profit to the company which supplies tasty tomatoes year round, garner 14 % of the total. Cotton and potato follow with 10 % and 8 %, respectively, while other crops are responsible for 11 % of the private sector's research.

The public institutions, with only 247 permits and notifications, have put their effort into the "other" category, with 38 % or a total of 93 research efforts. Tobacco, a popular laboratory research plant, is responsible for 22 % of research activities, while the more marketable crops of corn (3 %), potato (24 %) and tomato (13 %) round out public institutional research endeavors. On the face of it, there appears to be a distribution of research between public institutions and private industry in the preferred direction.

---

[2] Field tested organisms in the "other" category include: alfalfa, *Amelanchier laevis*, apple, *Arabicus thaliana*, barley, beet, belladonna, *Brassica oleracea*, carrot, chrysanthemum, *Cichorium intybus, Clavibacter*, cranberry, creeping bentgrass, *Cryphonectria parasi*, cucumber, eggplant, *Fusarium graminearum*, gladiolus, lettuce, melon, papaya, pea, peanut, pepper, petunia, plum, poplar, *Pseudomonas*, rapeseed, *Rhizobium*, rice, *Rubus idaeus*, squash, strawberry, sugarcane, sunflower, sweetgum, TMV, walnut, watermelon, wheat, and *Xanthomonas*.

Types of Experiments

Experiments being conducted[3] tend to reflect the different emphases of public institutions and private industry (see Figure 5). Private industry, with its need to quickly recover development costs and make a profit, is focusing on production characteristics that reduce production costs or add value to the product. Characteristics that reduce crop production costs, such as insect resistance, herbicide tolerance, and virus resistance, make up 64% of all research. Product quality interventions, which add value by enhancing the crops characteristics or creating new products, account for nearly a quarter of all experiments, followed by marker genes and the "other" category.[4]

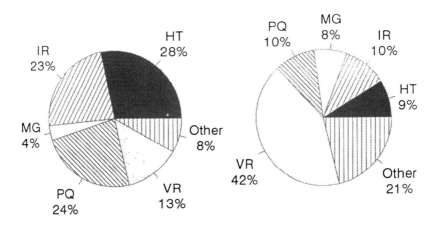

Figure 5: Experimentation by public and private institutions: Total APHIS field releases.

On the other hand, research by public institutions more closely reflects their scientific mission. Forty two percent of research is focused on virus resistance, an area of great scientific concern due to possible transfer of traits. The "other" category, with research on bacterial, fungicidal, and nematode resistance and other uses, such as the ability to grow in drought conditions and metabolize heavy metals and salts, is the next largest category (21 %). Insect resistance (10 %) and product quality (10 %) are next largest categories, followed by tests of

---

[3] Because more than one type of genetic intervention in the field testing of a crop can occur, the total number of experiments is greater than the number of field tests.

[4] The "other" category is made up of bacterial resistance, fungal resistance, nematode resistance, sterility, as well as other innovations that allow plants to grow in drought conditions and/or metabolize salts and heavy metals.

herbicide tolerance (9 %) and marker genes, which tend to be major areas of experimentation for private industry.

Confidential Business Information

The use of confidential business information (CBI)[5] did not occur until 1990, when six permits limited access to information. Since 1990, an increasingly greater proportion of field tests have one form or more of information classified in order to protect proprietary interests. Over this span, confidential business information has been invoked by both private industry and public institutions 55.8% of the time.

Private industry accounts for the great majority of field tests with confidential business information. Of a total of 1,941 permits and notifications, private industry has invoked CBI 1,008 times, compared to the 75 times it has been used by public organizations. In other words, public organizations have protected 30% of their research, while private industry has shielded 60 % of its research.

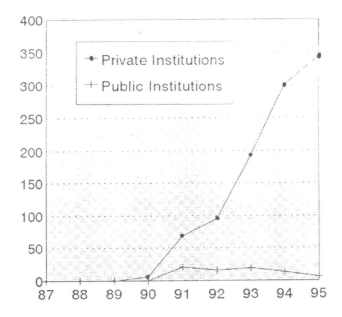

Figure 6: Public and private institutions with CBI: permits with confidential business information.

---

5 Confidential business information (CBI) refers to the classification of information so that it is not available to the general public. A major concern with CBI is balancing off the public's access to health and environmental safety data with the company's right to protect trade secrets and commercial and financial information that is privileged or confidential under the 1982 Freedom of Information Act (OTA, 1992:346-347).

Trends suggest that while the use of CBI by private industry has skyrocketed, public institutions are using it to a lesser extent (see Figure 6). Presumably, this means that private industry and public institutions are heading along two different paths, with private industry developing, and thus protecting, profitable technologies such as herbicide tolerant and pest resistant crops that are near commercialization. Meanwhile, public institutions are focused mainly on research dealing with basic exploratory science and products that may not offer market benefits.

Conclusions

Since 1987, agricultural biotechnology has expanded from a scientific approach to understanding crop growth to a technological endeavor with great economic and social promise. While scientific interest in agricultural biotechnology has remained steady, private industry interest in the possible economic benefits of plant products utilizing biotechnology has grown in an effort to share in what is seen as a US$50 billion industry by the year 2000 (Gore, 1994, 2). This is reflected in the phenomenal increase in permits and notifications for field release by private industry over the past six years.

While concern exists over the possible cooptation of university research agendas through increases in industry funding, these fears do not appear to have been realized. The field release data show that public sector research is not primarily focused on crops or experiments that have immediate economic value in the marketplace. There seems to be an implicit division of labor, with university and other publicly employed researchers focusing on areas not dominated by industry in both plant varieties and types of genetic intervention. Finally, the use of confidential business information by public institutions appears to be minor compared to its use by industry, assuaging some concerns that the overarching ethic of academic life, the free transfer of information, was in danger. In spite of this, its use by public institutions in 30 % of cases is cause for concern, and research into who uses it and how it is being used should be a priority.

Additionally, concerns over the direction of future research need to be addressed. Technology transfer centers, which infuse private industry funding into the academic research process, are increasingly becoming a part of the academic landscape. The mixture and/or competition of public and private sector funding sources may serve to obscure the ultimate constituency to be served, the general public.

The current emphasis on scientific advances that change the way research is done, and on enhancing competitiveness through technological innovation, raise questions about the proper role of public research institutions. Increased industry funding may solve short-term funding problems faced by the modern research university, but may also result in conflicting demands on the public sector researcher, i.e. the public good vs. private remuneration (Curry and Kenney, 1990; Knudson and Pray, 1991). In addition, it may serve to obscure potentially more effective, but less technologically advanced innovations such as integrated pest management (IPM).

While considering trends in agricultural biotechnology research is a first step in resolving these larger issues, it only serves to illuminate additional concerns. Agricultural biotechnology encompasses a vast range of policy issues that face the United States of America. An open dialogue with the public on the direction of agricultural research, the concerns it needs to address, and how we should fund the researchers who carry out that mission is needed.[6]

References

Biotechnology Project, The. (1993) Private Interests, Public Responsibilities, and the College of Agricultural and Life Sciences. White Paper for the Wisconsin Rural Development Center.

Biotech Reporter. (1995) Recent Partnerships in Bt Technologies. December, 3.

Biotech Reporter. (1996) Canada Approves Bt Corn. February, 3.

Blumenthal, D., Epstein, S. & Maxwell, J. (1986) Commercializing university research. *New England Journal of Medicine* **341**, 1621-1626.

Curry, J. & Kenney, M. (1990) Land-grant university-industry relationships in biotechnology: a comparison with the non-land-grant research universities. *Rural Sociology* **55**(1), 44-57.

Dorf, R.C. & Worthington, K.K.F. (1990) Technology transfer from universities and research laboratories. **37**, 251-266.

Dutton, G. (1994) USDA deregulates Calgene's bromoxynil-resistant cotton. *Genetic Engineering News* **14**, 1,16.

Etzkowitz, H. (1992) Individual investigators and their research groups. *Minerva* **30**, 28-50.

Etzkowitz, H. & Peters, L.S. (1991) Profiting from knowledge: organisational innovations and the evolution of economic norms. *Minerva* **29**, 133-166.

Falk, B.W. & Bruening, G. (1994) Will transgenic crops generate new viruses and new diseases? *Science* **263**, 1395-1396.

Gore, A. (1994) Government industry cooperation can help U.S. biotech succeed in global marketplace. *Genetic Engineering News* **14**, 2,12.

---

[6] The recent closings of the Office of Technology Assessment and the USDA's Office of Agricultural Biotechnology (OAB) severely hamper efforts to accomplish this, and place a greater burden on universities and interest groups and other organizations that represent the public.

Gray, D., Johnson, E.C. & Gidley, T.R. (1987) Industry-university projects and centers: an empirical comparison of two federally funded models fo cooperative science. *Evaluation Review* **10**, 776-793.

Hardy, R.W.F. (1986) Agricultural research in flux: experiments in university-industry ties. *Technology in Society* **8**, 273-276.

Holden, C. (1994) Tomato of tomorrow. *Science* **264**, 512-513.

Johnson, C. & Moore, R. (1990) American Universities, Technology Innovation, and Technology Transfer: Implications for Biotechnology Research. In: Biotechnology: Assessing Social Impacts and Policy Implications, Webber, D.J. (ed.). pp 55-68. Greenwood Press, Westport, CT.

Kenney, M. (1986) Biotechnology: the University-Industry Complex. Yale University Press. New Haven, CT.

Kleinman, D.L. & Kloppenburg, J. Jr. (1988) Biotechnology and university-industry relations: policy issues in research and the ownership of intellectual property at a land grant university. *Policy Studies Journal* **17**, 83-96.

Knudson, M.K. & Pray, C.E. (1991) Plant Variety Protection, Private Funding, and Public Sector Research Priorities. *American Journal of Agricultural Economics* August, 882-886.

Krimsky, S. (1991) Biotechnics and Society: The Rise of Industrial Genetics. Praeger Publishers. New York, NY.

Krimsky, S., Ennis, J.G. & Weissman, R. (1991) Academic-corporate ties in Biotechnology: a quantitative study. *Science, Technology, & Human Values* **16**, 275-287.

Lacy, W.B. & Busch, L. (1983) Informal scientific communication in the agricultural sciences. *Information Processing and Management* **19**, 193-202.

Lacy, W.B. & Busch, L. (1989) The Changing Division of Labor Between the University and Industry: The Case of Agricultural Biotechnology. In: *Biotechnology and the New Agricultural Revolution*, Molnar, J.J. & Kinnucan, H. (eds.) pp 21-50. Westview Press. Boulder, CO.

Lacy, W.B., Busch, L. & Cole, W.D. (1990) Biotechnology and Agricultural Cooperatives: Opportunities, Challenges, and Strategies for the Future. In: *Biotechnology: Assessing Social Impacts and Policy Implications*, Webber, D.J. (ed.). pp 55-68. Greenwood Press, Westport, CT.

Lynn, F.M., Poteat, P. & Palmer, B.L. (1988) The interplay of science, technology, and values in environmental applications of biotechnology. *Policy Studies Journal* **17**, 109-116.

MacKenzie, D.R., Jordan, J.P. & Clarke, N.C. (1992) Trends in Funding Biotechnology Research Through the Federal-State Partnership. White Paper for the USDA.

Marazita, C.F. (1991) Technology transfer in the United States: Industrial research at engineering research centers versus the technological needs of U.S. Industry. *Technological Forecasting and Social Change* **39**, 397-410.

Molnar, J.J. & Kinnucan, H. (1989) Biotechnology and the New Agricultural Revolution. Westview Press. Boulder, CO.

Office of Technology Assessment. (1992) A New Technological Era for American Agriculture. U.S. Government Printing Office. Washington, D.C.

Office of Technology Assessment. (1987) New Developments in Biotechnology-Background Paper: Public Perceptions of Biotechnology. U.S. Government Printing Office. Washington, D.C.

Postlewait, A., Parker, D.D. & Zilberman, D. (1993) The advent of biotechnology and technology transfer in agriculture. *Technological Forecasting and Social Change* **43**, 271-287.

Rogers, E.M. (1983) Diffusion of Innovations. The Free Press. New York, N.Y.

Ruscio, K.P. (1984) The Changing Context of Academic Science: University-industry relations in biotechnology and the public policy implications. *Policy Studies Review* **4**, 259-275.

Shelley, M.C. II, Woodman, W.F., Reichel, B.J. & Lasley, P. (1988) On the role of universities and biotechnology in economic development and public policy. *Policy Studies Journal* **17**, 156-168.

Shelley, M.C. II, Woodman,, W.F., Reichel, B.J. & Kinney, W.J. (1990) State legislators and economic development: University-industry relationships and the role of government in biotechnology. *Policy Studies Review* **9**, 455-470.

Souder, W.E. & Nassar, S. (1990) Choosing an R&D consortium. *Research-Technology Management* **33**, 35-41.

Webber, D.J. (1990) Biotechnology: Assessing Social Impacts and Policy Implications. Greenwood Press. Westport, CT.

Glossary

The following terms are defined in a general sense. Specific details are avoided to allow the general reader and the developing student to become literate in the language (and jargon) underlying molecular biology and genetics.

AFLP: amplification fragment length polymorphism. A variant DNA amplification product of different size produced by DAF, PCR, or RAPD. Recently this term has been applied to the high sensitivity method developed in the Netherlands by Dr. Marc Zabeau. The AFLP method of Keygene Inc. (sometimes called Key-PCR or SRFA) involves cutting DNA with two restriction nucleases, followed by ligation of an adapter DNA molecule, which in turn serves as a template for PCR primers, that are extended randomly on their 3' end, thereby giving selective restriction fragment amplification. Products are separated by polyacrylamide gel electrophoresis and give a complex display of bands. NB. the term AFLP was previously used by Caetano-Anollés et al (1991, Bio/Technology 9, 553-557) to describe amplification polymorphisms generated by the DAF method.

agarose gel: gel made up of seed weed derived agarose, used for separation of DNA molecules during electrophoresis. These gels are easily run, but lack the resolution of polyacrylamide gels. DNA is usually stained in agarose gels with the dye ethidium bromide.

Agrobacterium rhizogenes: bacterium causative for the hairy root disease, a form of organized tumor similar to the crown gall disease. Tumors also contain bacterial DNA transferred and covalently integrated in the plant genome. This species can be used to generate chimeric plants with a transformed root system and a untransformed shoot.

Agrobacterium tumefaciens: bacterium causative for the crown gall disease capable of tumor induction in plants. This is achieved through the transfer of a region of DNA (T-DNA) which codes for the synthesis of plant growth regulators.

allele: one of the forms of a gene.

amino acid: a small organic molecule which is a building stone for proteins or polypeptides. There are 20 amino acids used in biological systems, although chemically speaking there are many more. Some are essential for humans, as their synthesis is not possible in *Homo sapiens*.

amplicon: DNA region defined by two opposing primer amplification sites.

aneuploid: chromosomal number is more or less than the diploid number by sets lesser than the haploid number (e.g., humans with 47 or 45 chromosomes are aneuploids).

antibiotic: substance which acts to destroy or inhibit the growth of a microbe (e.g., bacteria or fungi).

antibody: common name for an immunoglobulin protein molecule which reacts with a specific antigen.

antigen: foreign molecule recognized by the immune response system of an animal.

antisense: strand of DNA or RNA equivalent to the antisense strand of a gene. Sense and antisense strands will anneal *in vivo* and cause inactivation of the gene's expression.

AP-PCR: single primer amplification method developed by Welsh and McClelland (1990). Usually large single primers of 20 to 30 nt length are annealed to target DNA then amplified by PCR under non-stringent conditions for 2 cycles. Then stringency (i.e., temperature is increased). Products are separated by polyacrylamide gel electrophoresis and autoradiography.

ATP: adenosine triphosphate. The general energy donor molecule in all organisms.

autoradiography: method used to detect radioactive substances by the property to darken film superimposed on the compounds. Can be used on whole organisms or molecules separated by molecular methods such as electrophoresis.

auxin: plant hormone involved in cell elongation and growth (e.g., indole acetic acid, 2,4-D).

bacteriophage: bacterial virus that either kills or infects latently a bacterium. Often abbreviated as 'phage'. Phage lambda (λ) is most studied, as is frequently used for cloning experiments.

bacterium: a relatively simple, single celled organism. All bacteria are prokaryotes; therefore they lack a defined nucleus and chromosomes as characteristic of eukaryotes such as yeast, plants and animals. The best studied bacterium is *Escherichia coli* (*E. coli*), which inhabits the digestive tract of many animals.

bacteroid: the nitrogen-fixing form of a symbiosome-contained *Rhizobium* or *Bradyrhizobium* bacterium found inside nodules.

basepair: two nucleotides hydrogen-bonded according to Chargaff's rules (A = T; G = C).

biolistics: process by which DNA molecules are propelled into a recipient cell using coated microprojectiles shot from a 'gene gun'. The method of propulsion may vary and ranges from electric discharge to helium blast.

biotechnology: the combination of biochemistry, genetics, microbiology, and engineering to develop products and organisms of commercial value.

Bradyrhizobium: bacteria able to form nodules and fix nitrogen in association with some legumes such as cowpea, soybean and peanuts.

BSA: bulk segregant analysis. Detection strategy used to define molecular regions controlling a certain phenotype. Large populations of organisms are usually pooled (bulked) into two classes representing distinguishing phenotypes. DNA profiling by DAF or RAPD is used to detect distinguishing band patterns.

BT toxin: *Bacillus thuringiensis* protein crystal, which if ingested by certain insects will cause death. The gene is commonly used in plant transformation to achieve insect resistance.

CaMV: cauliflower mosaic virus. One of the few DNA plant virus. Its 35S promoter is often used for transformation experiments.

cathode: negatively charged electrode in electrophoresis. (as compared to anode, the positively charged electrode).

cDNA: complementary DNA; DNA made by reverse transcriptase enzyme from RNA.

cell cycle: the progression of different stages of DNA replication and cell division in eukaryotes. DNA replication is called S-phase (synthesis) while cell division is called M-phase (mitosis). S-phase and M are separated by the Gap 2 (G2) phase, while M-phase is followed by Gap 1 (G1) phase (cf., MPF, cyclin B).

centiMorgan (cM): unit of recombination. Named after the geneticist Morgan. One cM is equivalent to one percent recombination.

centromere: chromosomal region functioning as the spindle attachment region to allow chromosome and/or chromatid separation during mitosis and meiosis.

Chargaff's rules: stipulate that since in double stranded DNA the amount of adenine (A) equals that of thymine (T) and guanine (G) that of cytosine (C). Accordingly A binds to T and G to C by hydrogen bonds, giving the DNA molecule the properties needed for replication and information storage.

chloroplast: site of photosynthesis in eukaryotes. Contains circular DNA.

chromatin: complex form of eukaryotic nuclear material at the times between cellular divisions (cf., euchromatin, heterochromatin).

chromosome: organized structure made up of DNA and proteins. Visible through light microscopy during cell division (cf., mitosis and meiosis).

chromosome walking: strategy of chromosome analysis in which cloned chromosomal segments are used to isolate sequentially neighboring DNA to generate a series of overlapping clones, called a contig. Eventually this contig will include the region of the desired gene. Walking like this has disadvantages in eukaryotes, because of highly repeated DNA.

clone: Greek word 'klon' meaning 'twig'. A method of vegetative reproduction of an organism. Commonly used in horticulture as cuttings or drafts. Resulting organisms are defined as a clone meaning they were derived from the same original source organism. Commonly clones are presumed to be genetically identical. This may not be the case because of further genetic change after the original duplications. In modern genetic jargon, clone describes an isolated DNA sequence ligated into a bacterial plasmid or virus, so that the sequence can be propagated indefinitely using microbiological means. Such cloned sequence can be used for sequencing, expression studies, or as probe.

CMS: cytoplasmic male sterility caused by alterations in mitochondrial DNA.

co-dominant: both alleles at the locus influence the phenotype of the heterozygote resulting in a new phenotypic class.

codon: arrangement of three nucleotides in mRNA controlling the insertion of an amino acid into a polypeptide.

contig: contiguous fragment of DNA used in genome analysis.

cortex: bulk tissue of a plant root or legume nodule. Characterized by vacuolated cells and absence of mitotic divisions.

cyanobacterium: a prokaryote, commonly a blue-green alga. Capable of photosynthesis and sometimes of nitrogen fixation.

cytokinin: plant hormone involved in cell division and senescence (e.g., kinetin, zeatin).

DAF: DNA amplification fingerprinting. A patented method of general DNA amplification (cf., PCR) using a single primer of between 5 and 20 nucleotides in length (8-mers are optimal). Primer concentrations are usually much higher and template concentrations lower than in the RAPD reaction. The primer is arbitrarily chosen and may generate as many as 80 amplification products, which can be resolved by a variety of methods, including polyacrylamide gel electrophoresis (PAGE) and silver staining, agarose electrophoresis, and automated analysis using either DNA sequencers (using fluorescent primers and laser detection) or capillary chromatography. Method was developed in 1990 by the University of Tennessee and is used to distinguish organisms (e.g., cultivar of one crop species), gene mapping, and diagnostics (see MAAP, RAPD and AP-PCR).

DGGE: denaturing gradient gel electrophoresis. Method to separate DNA or proteins by altering the pH inside a gel. As the molecules migrate in the electric field, they encounter denaturing conditions, which slows their mobility. Different molecules respond differently to the denaturation condition and therefore respond at a different severity of denaturant (resulting in a loss of mobility at a different point in the gel).

differential display: also called DD-RT-PCR. Method used to distinguish mRNA populations by first converting the mRNA to cDNA using reverse transcriptase (RT), then using a PCR amplification with two primers, one arbitrary as described for DAF or RAPD, the other anchored on the polyA tail of the original message. Products are usually resolved by sequencing gels and autoradiography.

diploid: the normal somatic chromosome number of an organism (twice haploid).

DNA: deoxyribonucleic acid. The genetic molecule of most organisms, except some viruses. Double stranded polymer of nucleotides arranged along a deoxyribose and phosphate backbone. Structure was proposed by James Watson and Francis Crick in 1953.

dominant: form of expression of a gene, in which the phenotype of the dominant form is expressed over the recessive form.

electrophoresis: method used to separate protein or nucleic acid molecules in an electric field extending across a physical medium such as agarose gel or polyacrylamide. Smaller molecules travel faster and therefore are found further down the gel. Since DNA is negatively charged (because of the phosphate), the cathode is at the upper end. Derived from Greek: phoro, meaning I carry.

endodermis: cell layer separating cortex and pericycle in plant roots. Contains the Casperian Strip, which is important in diffusion resistance and turgor relationships in plants.

enzyme: a biological catalyst allowing the completion of biochemical reactions. Most enzymes are proteins although some RNA enzymes were recently discovered (see ribozyme).

epidermis: external cell layer of a plant.

epistasis: mode of gene interaction where the expression of one gene influences the expression of another at another location in the genome. For example, non-nodulation genes in soybean epistatically suppress supernodulation.

Escherichia coli: also *E. coli*. A common gut bacterium used as a model genetic organism. *E. coli* has about 3000 genes and a genome of around 4 million basepairs.

ethidium bromide: chemical used to visualize DNA by fluorescence (see agarose gels). Interpositions itself into the DNA groove and alters buoyancy.

ethylene: gaseous plant hormone involved in stress responses, fruit ripening as well as nodulation of legumes.

euchromatin: the portion of genomic DNA which remains relatively unstained and is transcriptionally active.

eukaryote: organism characterized by the presence of a nucleus. Also other organelles such as mitochondria and/or chloroplasts may be present in eukaryotes. Includes all plants, animals, green algae and fungi.

exon: expressed region of a gene. Transcribed and translated.

F_2: filia 2; the second generation in a hybridization being the product of the F_1.

FISH: fluorescent *in situ* hybridization.

flavone: aromatic molecule (i.e., contains a benzene ring as a core molecule) significant in the communication of legume plant to *Rhizobium* and *Bradyrhizobium*.

fluorescence: light produced by substance after external stimulation with another light (or energy) source at another energy level.

Frankia: an actinomycetes bacterium capable of forming nodules with some non-legumes such as *Alnus* and *Casuarina* (tree species).

gamete: the haploid sex cell, such as a pollen grain, a sperm or an egg.

gene: functional unit of inheritance. Usually a gene is defined as that region of DNA that controls the synthesis of a polypeptide.

genetic code: the conversion cipher which allows the interpretation of triplet codons to their matching amino acids.

gene probe: a specific single stranded DNA sequence, often labeled with radioactive phosphorous (^{32}P) to help the detection of a gene through the complementary (see Chargaff's rules) sequence to which the probe will bind. Hybridization is usually detected by autoradiography.

genetic engineering: the directed genetic manipulation of an organism using recombinant DNA molecules not commonly found in nature.

genetic fingerprinting: a method probably initially developed by Alec Jeffreys which enables genetic relationships between close relatives to be established using DNA technologies.

genome: the entire set of hereditary molecules in an organism.

genotype: the genetic make-up of an organism which, depending on the environment and other genes, may be expressed as the phenotype.

GISH: genome *in situ* hybridization.

glyphosate: the active ingredient in the herbicide Round-Up.

gus: abbreviation for β-glucuronidase, an enzyme that produces a blue colored pigment from a colorless substrate. The gene coding for the enzyme is frequently used as a reporter gene in gene transfer experiments.

hairpin: DNA or RNA structure formed by self-annealing forming a stem and loop or hair-pin.

haploid: the gametic chromosome number of an organism, as found in sex cells.

herbicide: substance used to kill plants. Usually used to eliminate weeds. Many herbicides now are applied in small volumes and are biologically non-persistent. Old herbicides like atrizine accumulate in the environment, while new types, like Round-Up and BASTA (Ignite) degrade radidly and have no reported effect on animals.

heterochromatin: portion of genomic DNA which is highly stained because of high condensation. The heterochromatin is transcriptionally inactive and often contains large amounts of highly repeated DNA (satellite DNA). Heterochromatin comes in two forms: facultative heterochromatin will alter in staining and activity at different developmental stages (i.e., the inactivated X-chromosome in mammalian females). Constitutive heterochromatin is constantly "turned off". Often it is found around the centromere.

heterosis: also called hybrid vigor. Interaction of multiple genes causes an improvement of the heterozygote above the level of both parents.

homeologous: related chromosome in a polyploid organism, which shares perhaps a common ancestry, but may contain different genes and alleles. For example, wheat is a hexaploid and carries three chromosome sets of 2. Within each set the chromosomes are homologous, between the sets they are homeologous.

homogenotization: bacterial genetics procedure used to exchange genetic markers from a plasmid to the recipient linkage group. Also called marker exchange.

homologous: partner chromosome in a diploid organism. Generally the homologous chromosome is similar as it carries the same genes, but is different as many allelic differences may exist.

hormone: regulatory substance which acts at low concentrations (less than one micromolar) and at a distance from its site of synthesis. Controls metabolism and development.

in situ hybridization: cytological method, which couples molecular biology, biochemistry and cytology. Usually the expression of a gene (as RNA) or the location of target DNA (for example in a chromosome) is detected by a molecular probe. The probe can be labeled by a variety of biochemical methods.

in situ: Latin: in the natural location.

in vitro: Latin: meaning: in glass (i.e., inside a test tube).

in vivo: Latin: meaning : in real life.

intron: intervening sequence in genes. Transcribed but not translated.

isoflavone: aromatic signal substance involved in the nodulation of legumes.

isozyme: also called allozyme. An enzyme which exists in multiple electrophoretic forms either because of allelic variation within one polypeptide or multimeric associations of variant forms.

karyotype: the pattern and shape of chromosomes of an organism.

kilobase: kb; one thousand basepairs of DNA.

legume: plant family characterized by a pea-like flower morphology. Many but not all legumes are nodulated and form nitrogen-fixing symbioses with soil bacteria called *Rhizobium, Bradyrhizobium,* and *Azorhizobium.*

ligase: an enzyme used to couple double-stranded DNA molecules together.

locus: the chromosomal position of a genetic condition as defined by a detectable phenotype.

Lotus japonicus: a legume used for forage. Because of its experimental advantages such as low genome size (N = 6), the plant is considered as a model for legume research, capable of high transformation with *Agrobacterium*.

MAAP: multiple arbitrary amplicon profiling. A collective term used to describe the RAPD, AP-PCR and DAF reactions.

map: the ordered arrangement of genes or molecular markers of an organism, indicating the position and distance between the markers and loci. Most maps are genetic maps based on the percentage recombination. Some maps are cytological maps based on the arrangement of chromosomal regions, while others are physical maps based on the amount of DNA between markers and loci.

MAPMAKER: a computer program which allows the ordering and mapping of molecular and genetic markers. Others exist as well.

marker: a distinguishing feature that can be used to identify a particular part or region of a chromosome or genetic linkage group.

megabase: Mb; one million basepairs of DNA.

meiosis: cell division in eukaryotes giving rise to gametes of different genetic make-up to the parental cell and to each other.

meristem: organized zone of mitotic division giving rise to cell clusters capable of further differentiation into new organ types.

messenger RNA: mRNA; product from DNA by transcription which serves as the information carrier for translation in proteins.

methylation: enzymatic addition of a methyl (CH_3-) group to DNA which causes inactivation of that region. Usually CpG nucleotide pairs are target for this addition.

microtubule: cellular strand needed for chromosome separation as well as cytoplasmic shape and targeting of biochemical processes. Composed mainly of a protein called tubulin.

mis-matching: error during annealing of primer to template or probe to target, caused by low stringency.

mitochondrion: organelle found in all eukaryotes. Site of respiration (ATP synthesis). Contains its own DNA.

mitosis: cell division in eukaryotes giving rise to two genetically identical progeny cells. Occurs frequently in meristems.

molecular marker: a molecular signpost used in eukaryotic gene isolation. Usually a RFLP probe or a primer site for DNA amplification.

mutant: an organism or gene with inheritable altered phenotype from the wild type.

NIL: near-isogenic lines. Used frequently for defining molecular regions of a genome that cause a phenotype. Also of value in agronomy and plant breeding.

nitrogen fixation: process by which nitrogen gas is converted to ammonia. This process occurs frequently in bacterial induced nodules of legumes and results in an independence on fertilizer nitrogen. Also possible by the industrial Haber-Bosch process.

nodule: outgrowth from the roots (or stems in some cases) of legumes induced by bacteria or exogenous agents such as bacterial derived nodulation factors or auxin transport inhibitors.

Northern hybridization: method to detect RNA by use of a probe (cf., Southern hybridization).

nucleotide: component of DNA and RNA; also known as 'base'. Two nucleotides paired according to Chargaff's rules are one base pair.

nucleus: organelle found in cells of all eukaryotes. Contains chromatin and is membrane-bound.

oligonucleotide: a polymer of nucleotides usually 5 to 30 base pairs long.

organelle: membrane bound cellular compartment (e.g., nucleus, chloroplast, mitochondrion, Golgi apparatus).

PCR: Polymerase Chain Reaction. A method for amplifying DNA of any organism using two specific oligonucleotide primers (about 15 base pairs in length) which flank the region of interest. The method was developed by CETUS Corporation in the mid-1980s and is of extreme value in diagnostics, forensics and general molecular biology (e.g., sequencing, probe preparation, genome mapping). Commercial rights are now owned by Hoffman-Roche; inventor K. Mullis received the Nobel Prize in 1993.

peribacteroid unit: see symbiosome.

pericycle: cell layer surrounding the vascular bundle in plant roots. Gives rise to lateral roots as well as nodules in actinorhizal plants such as Alnus. Is also involved in legume nodule formation.

PFGE: see pulse field gel electrophoresis.

phenotype: the appearance of an organism taken as a genetic characteristic.

phosphorylation: biological process by which proteins are 'decorated' with phosphate groups derived from ATP. The process alters the biological activity of the protein whereby facilitating a form of physiological regulation.

phytoalexin: substance involved in the antimicrobial response of a plant.

plant growth regulator: broad class of chemicals that control the growth of plants. Many are also natural compounds found within plants, where they may act as hormones.

plasmid: circular, covalently closed DNA molecule commonly found in bacteria. Often used as a cloning vector in genetic engineering.

plastid: generic name for all chloroplast-like organelles found in different plant cells.

pleiotropy: mode of gene expression where the expression of one gene has multiple phenotypes.

polymorphism: the presence of several forms of a genetic characteristic in a population. Molecular polymorphisms are caused by alterations in the DNA sequence as detected through sequencing or DNA fingerprinting.

polyploid: more than diploid by multiples of the haploid number. For example, banana is a triploid. It is polyploid.

polysaccharide: polymer molecule of sugars (e.g., starch, glycogen).

positional cloning: experimental approach used to locate and isolate gene sequences for which the gene product is not known. Instead the phenotype is mapped and large fragments are isolated in the region of informative molecular markers known to segregate closely with the gene of interest.

primer: short sequence of DNA (or RNA) used to initiate DNA replication.

probe: a known sequence of DNA (or RNA) used to detect homologous sequences in DNA or RNA after reassociation based on Chargaff's rules.

prokaryote: a bacterium. Characterized by the absence of major organelles such as the nucleus and plastids.

promoter: regulatory region of a gene involved in the control of RNA polymerase binding to the target gene.

protein: a polymer of amino acids usually with structural roles (such as keratin, the hair protein) or catalytic roles (see enzyme).

pulse field gel electrophoresis: PFGE; a variation on the electrophoresis procedure insofar that a computer flips the electric field in preset pulses in different directions and defined strengths. Method is used to isolate very large DNA molecules (greater than one megabase; million basepairs) including whole chromosomes of organisms such as yeast.

PVR: plant variety rights. Legislation that exists in many countries which allows a copyright to be obtained for a specific plant.

QTL: quantitative trait locus; a term given to a genomic region which controls a phenotype by interaction with other genes (i.e., oil content in soybean).

RAC: random access chromatography. Method developed by Leon Whaler for quick analysis of polymorphic forms, avoiding use of priority listing.

RAPD: random amplified polymorphic DNA. A widely used DNA amplification method developed by DuPont Company related to DAF, as it uses single primers. Method is restricted to primers 9 nucleotides and larger, annealing temperatures below 45°C, and amplification products are generally visualized using agarose electrophoresis and ethidium bromide fluorescence. Usually about 4 to 10 products are generated. Method is useful for mapping approaches.

rDNA: recombinant DNA; made by the joining of DNA fragments from different species using restriction endonuclease and cloning approaches.

receptor: protein molecule, usually on the cell surface, able to receive and interpret an external signal. Some receptors are on the nuclear membrane or in the nucleus.

recessive: form of expression of a gene, in which the phenotype of the recessive form is not expressed over the dominant form.

recombination: natural process of exchanging DNA fragments between different DNA molecules. Occurs in both prokaryotes and eukaryotes, but by slightly different processes. Eukaryotic recombination occurs predominantly during meiosis and gives rise to gametes of non-parental gene combinations.

restriction endonuclease: an enzyme which cuts (or restricts) DNA at specific sequences generating fragments (e.g., *Eco*RI).

reverse transcriptase: enzyme able to synthesize DNA from RNA. Often found in tumor viruses.

RFLP: restriction fragment length polymorphism. DNA fragment difference generated by the action of a restriction endonuclease on the DNA of two or more organisms and detected usually by Southern hybridization using a radioactively (or chemically) labeled probe. Modern detection is possible by a phospho-imaging apparatus.

Rhizobium: bacteria able to form nodules with some legumes such as peas, alfalfa, and clovers.

ribosome: site of protein synthesis in prokaryotes and eukaryotes. Made up of two subunits, comprising three RNA molecules and about 50 proteins.

ribozyme: enzyme made entirely of RNA.

RIL: recombinant inbred lines. Used for mapping purposes. These are the product of an initial cross between two parent lines and the subsequent selfing of the F_2 individuals.

RNA: ribonucleic acid. Used as messenger RNA, transfer RNA and ribosomal RNA. Some RNA molecules are the genetic molecule of viruses (e.g., tobacco mosaic virus and HIV) and viroids. Some RNA molecules may have enzymatic activity (see ribozymes).

rol **genes**: set of genes from *Agrobacterium rhizogenes* which cause the formation of transformed roots on a large variety of dicotyledonous plants. These genes seem to accentuate the normal plant hormone levels.

root hair: protruding from an epidermal cell. Grows by tip elongation.

SCAR: sequence characterized amplified region. A useful molecular marker which may be generated by the partial sequencing of AP-PCR, DAF, or RAPD band or an RFLP clone. Resultant PCR primers then provide specificity.

silver staining: a procedure by which DNA, proteins or polysaccharides are visualized using silver complexes.

Southern hybridization: also Southern blotting. Method employing gel separation of restricted DNA fragments, their blotting onto a membrane support, dissociation into single stranded DNA and hybridization (reassociation) with a labeled probe. Regions of homology are detected usually by autoradiography. Invented by E. Southern in 1975.

soybean: *Glycine max* (L) Merr. A tropical legume of wide agronomic application. Nodulates with *Bradyrhizobium japonicum* and *Rhizobium fredii*.

SRFA: selective restriction fragment amplification. DNA profiling method developed by KeyGene Inc. in the Netherlands using a combination of PCR, adapter ligation and DNA restriction. (see AFLP).

SSR: single sequence repeat. Also called microsatellite. These are repeats of two or three nucleotides, sometimes found up to 30 times repeated. External PCR primers may detect length variation, which can be mapped.

Stoffel fragment: truncated *Taq* polymerase without its exonuclease activity. Used in PCR, RAPD and DAF. Named after the investigator Dr. E. Stoffel.

STR: short tandem repeat; same as microsatellite or SSR.

stringency: term used to define the accuracy of nucleic acid interactions in Southern or Northern hybridization experiments. Low stringency permits errors (or mismatches). High stringency, usually attained by elevated temperature, means high fidelity annealing. Term is also applicable to primer-template interactions.

stroma: central region of a chloroplast filled with thylakoid membranes.

STS: sequence tagged site. Region of DNA on a chromosome used as a signpost for molecular gene mapping approaches.

supernatant: liquid left above a pellet after centrifugation. Derived from Latin: super-natare: to swim above.

symbiosis: mutually beneficial living together of two organisms of different species (e.g., nodulation in legumes).

symbiosome: organelle structure found in nodules of legumes. Encases the symbiotic form of *Rhizobium* and *Bradyrhizobium* called the bacteroid. New term for 'peribacteroid unit (PBU)'.

Taq **polymerase**: thermostable DNA polymerase from *Thermus aquaticus*.

telomere: terminal region of chromosomes characterized by repeated DNA sequences.

thermocycler: an apparatus which changes temperatures according to precise programming. Usually used with PCR, RAPD, AP-PCR, AFLP and DAF.

Thermus aquaticus: thermophylic bacterium found in hot springs. Its DNA polymerase enzyme is thermostable and is used in the PCR, RAPD, and DAF.

transcription: the process by which DNA is copied into RNA. As the nucleic acid 'language' stays the same (see genetic code), the process is called transcription (cf., translation).

transformation: if used in a genetic sense, this term implies the transfer of a gene to a new cellular environment, coupled with the expression of that gene to alter the phenotype of the recipient cell. May be transient or stable as judged by inheritance.

transgenic: an organism containing genetic material artificially placed there from another organism. Usually transgenics involve interspecies transfer.

translation: the process by which RNA is made into proteins. Occurs in ribosomes. Called translation because the nucleic acid 'language' based on a sequence of four 'letters' arranged in triplet codons is changed to a 20 component 'language' used in protein synthesis.

transposon: transposable element in either prokaryote or eukaryote. May contain its own transposase enzyme gene. Usually is flanked by direct or indirect repeat sequences. Transposons are useful genetic tools as they inactivate gene upon insertion.

trisomic: three chromosomes of one type are present within a genome.

vector: a plasmid, virus or other vehicle used to carry an isolated DNA sequence from one cell to another. Usually involved in the production of interspecific transgenic organisms.

virus: a sub-microscopic agent which contains genetic material (either DNA or RNA, but never both), but must invade another living cell to replicate itself.

VNTR: variable number tandem repeat. Also called mini-satellite. Used in DNA fingerprinting.

Western hybridization: also called Western blotting. Method to detect proteins by use of an antibody directed against it.

wild type: the normal form of an organism (i.e., not mutant). Note: this is a noun (not hyphenated). c.f., wild-type, which is the adjective.

YAC: yeast artificial chromosome. Used extensively in the cloning of large DNA molecules used in eukaryotic gene mapping.

Index

A

ABA 31
ADP-ribosylation 68
ablation 70, 80, 81
acetic acid 104
acridine orange 179
acrolein 75
Aegilops 149, 161
AFLP 114, 134, 145
Agricola 4
agricultural biotechnology 192, 194, 202
Agrobacterium 67, 82
Agrobacterium rhizogenes 76
Agrobacterium tumefaciens 75
agropine 76
Agropyron 149, 161
allergen 17
Alternaria solani 29
aminoglycoside 3'-phosphotransferase II 21
amphidiploid 154
amplicon 133, 164
amplification 158
anionic peroxidase 36
anther 70
anther box 86
anti-cancer drug 106
antisense 14, 59, 67, 71, 73, 79, 83, 192
AP-PCR 145
APH(3')II 17
aphid 59
APHIS 6, 17, 196
Apium graveolens 123
apoplast 86
apple 118
Arabidopsis 28, 33, 71, 75, 76, 81, 85, 86, 124, 171
ascocarp 43
asepsis 105
Asgrow Seed 193
Aspergillus oryzae 70
Automation 143
aux2 gene 76
avirulence 27
avrPto 27
Azolla 143

B

β-aminobutyric acid 32
β-glucuronidase 69
β-1,3-glucan 28
β-1,3-glucanase 27, 28
β-glucuronidase 83
β-phaseolin 82
Bacillus amyloliquefaciens 70
Bacillus thuringiensis 193
banana 120
barley 149, 173
barnase 81
barstar 86
bermudagrass 140, 143
BIC 4
biocontrol 41, 43, 46
bioluminescence 34
black root rot 42
blueberry 121
Brassica campestris 83
Brassica napus 69, 81
breast cancer 106
broad-spectrum resistance 37
broccoli 122
bromoxynil 17
budworm 17
BYMV 59

C

cabbage 34, 76
callose 27, 188
CaMV 75, 86
capillary electrophoresis 137
Carica papaya 119
carnation 27
castor bean 70
catalytic domain 72
cauliflower 122
celery 122
cell death 36
cell wall 30, 167
cell wall hydrolases 27
Cf9 gene 28
CFSAN 21
chalcone synthase 86
challenge virus 60
chemotaxis 167
chitin 28
chitinase 27, 29
chloramphenicol acetyltransferase 72
chloroplast 71, 72, 79, 150
citrate synthase 83
Citrus 118

Cladosporium fulvum 28, 123
clearance 192
cleavage product 72, 80
cloth 185
CLSM 177
coat protein 51
cocoa 119
commercialization 98
confidential business information 201
confocal microscopy 188
corn 137, 192, 197
costs 200
cosuppression 60, 67, 74
cotton 19, 143, 178, 197
cotton fiber 185
CRADAs 3
criminology 141
CRIS 4
crop 197
CSREES 6
cucumber 27
Cucumis melo 53
cutin 188
cytochrome P450 86
cytokinin 77
cytosine deaminase 78
cytotoxin 67

D
DAF 113, 122, 124, 132
database 1
DGGE 137
2,4-diacetylphloroglucinol 41
dianthus 124
differential display 135
Dinkel 154
diphtheria toxin 68
Diplodia maydis 43
disease control 25
DNA methylation 76
DNA polymerase 144
DNA synthesis 78
dogwood 143
drought tolerance 149

E
early ripening 149
Einkorn 154
elongation factor EF-2 68

Elymus 149
Elytrigia 149, 163
embryogenesis 83
Emmer 154
epifluorescence 177
ethephon 31
ethionine 170
European Community 193
European corn borer 193

F
FDA 17
Federal Government 1
fenthion 75
fertility 81
field release 195
field trial 16
flavor 15, 102
FLAVR SAVR 13, 192
flower 83
fluorescein diacetate 179
fluorescence 183
fluorescent pseudomonad 41
fluorochrome 188
fluorophenylalanine 171
5-fluorocytosine 78
5-fluoroindole 75
5-fluorouracil 78
forensics 131
Fragaria x ananassa 120
freeze-drying 99
fruit trees 112
fungal pathogens 16
Fusarium oxysporum 27, 31

G
γ1-tubulin 82
Gaeumannomyces graminis 41
ganciclovir 77, 84
geminivirus 71
GENBANK 3
gene inactivation 76
Gene Shears 72
gene targeting 84
Gene-for-gene 27
genetic maps 111
Geotrichum candidum 16
gerber daisy 124
glutamate semialdehyde aminotransferase 71

Glycine max 139
glycoalkaloids 20
glycosylation 18
GMENDEL (program) 115
Gossypium hirsutum 178
governmental institutions 197
grape 120
Growmark 193
growth regulator 168
GUS 60

H
hammerhead 71
health food 102
heat shock 77
herbicide 17, 168
herbicide tolerance 200
homologous recombination 79
Hordeum 150, 162
horizontal gene transfer 18
horticulture 112
HSV 77
human 18, 174
hypersensitive reaction 27, 87

I
IAA 31, 75
indole oxidase 72
indole-3-acetamide 75
indole-3-acetamide hydrolase 75
induced systemic resistance 25
industry 194
intellectual property 3
Internet 10
isozyme 30

J
JOINMAP 115

K
kanamycin 18, 77, 79, 80, 192
kinetin 31

L
large-scale planting 192
lettuce 122
leucine-rich repeat 28

Leymus 150, 161
lignification 27, 36
linkage 112
Lotus japonicus 78, 143
lysozyme 29, 34

M
MAAP 132
MacGregor's® 21
maize 141, 149
male fertility 86
male sterility 71, 81
male-sterile 69
Malus domestica 115, 117
Mangifera indica 119
mango 119
map 139
map-based cloning 28
MAPMAKER 115
marker-assisted breeding 137
marker-assisted selection 111
market needs 96
Maximizer™ hybrid corn 193
Meloidogyne incognita 117
Mendelian 19
metalloprotease 46
5-methyl-tryptophan 75
microgametophyte 178
microsatellites 114, 135
mini-hairpin primer 140
mismatch 135
Mycogen Plant Sciences 193

N
N locus 28
N-dealkylation 79
nandina 143
napin 69
NBIAP 6
NCBI BLAST 168
near-isogenic lines 140
negative selection 67, 71, 73
nematode 139
neomycin 19
neomycin phosphotransferase 80
Neverripe tomato 16
New Leaf™ potato 193
Nicotiana benthamiana 86

Nicotiana plumbaginifolia 73
Nicotiana tabacum 53, 59, 78
nitrate 168
nitrate reductase 75
nonhomologous recombination 84
nos promoter 79
NRICGP 7
NTIS 8
nucleotide binding site 28

O

Oleo europaea 119
oligonucleotide 144
olive 119
onion 123
ornamentals 112
overexpression 67
oxylysine 170

P

p-hydroxybenzaldehyde 104
p-hydroxybenzoic acid 104
p-hydroxybenzyl methyl ether 104
P450$_{SU1}$ monooxygenase 79
papaya 119
partnerships 192
paternity 141
pathogen 139, 174, 188
pathogenesis 28
PAUP 159
PCR 113, 118, 131, 144, 158
pea 123
peach 116, 124
pectin 19
peptidoglycan 28
permits 16, 198
peroxidase 27, 37
petunia 76, 124, 156
PGPR 26
pharmaceutical 107
phaseolin 83
phaseolotoxin 174
phenazine-1-carboxylate 41
phosphoglycerate kinase 171
phylogenetic 141, 153
phytoalexin 30
phytochrome 85
phytoene synthase 71
phytopathogens 25

Phytophthora infestans 87, 114, 122
Phytophthora parasitica 30
phytotoxicity 79
PhytoVanilla 95
pistil 81
Pisum sativum 123
pleiotropic 19
pokeweed 86
pokeweed antiviral protein 70
pollen 81, 183
polygalacturonase 13
polymorphism 111, 140, 158
potato 86, 87, 114, 121, 197
potyvirus 51, 58
PPV 53
PR-1 protein 25, 28
preproricin 86
primer 119, 140, 159
private industry 195, 201
pro-herbicide 79
product 103, 139, 144, 194
product applications 100
Profiling-by-Hybridization 144
promoter 68, 74, 171
protection 59
proteinase 55
Prunus persica 115
Psathyrostachys 150
Pseudomonas aeruginosa 69
Pseudomonas fluorescens 27
Pseudomonas phaseolicola 174
Pseudomonas syringae 27, 75
Pseudomonas tabaci 174
Pto gene 27, 75
public institutions 201
PVX 60
PVY 53, 59
pyoverdine 46
Pyrenophora tritici 43
Pythium ultimum 42

Q

QTL 122

R

RAPD 113, 116, 124, 132, 145
rapeseed 17
ras2 gene 73
raspberry 121

RDRP 55, 61
recovery 57
research trends 191
resistance 19, 31, 51, 193
resistance gene 149
RFLP 114, 117, 121, 124, 140, 150, 163
rhizobacteria 26
Rhizoctonia solani 37
Rhizopus stolonifer 16
rhizosphere 45
rhodamine phalloidin 179
Rhodococcus 72
ribonuclease 80
ribosome 80
ribotoxin 70
ribozyme 72, 80
rice 149
ricin 70, 86
Ricinus communis 70
RIP 70, 86
ripening 14
risks 191
RNA 150
RNA packaging 60
RNA polymerase 55
RNA replication 55
RNase 70
robotics 124
rolC gene 86
Round-Up 192
rps2 gene 28

S

Saccharomyces cerevisiae 172
salicylic acid 28, 31, 45
Saponaria officinalis 70
SAR 25, 45
SBIR 4, 8
scale-up 96
SCAR 113, 122, 137
Sclerotinia sclerotiorum 43
Secale 150, 162
self-incompatibility 81
self-sterility 81
sequence similarity 55
shelf life 14
signal transduction 28, 77, 85
silver staining 113, 121, 135

slime mold 73
snapdragon 124
softening 14
solanine 20
Solanum tuberosum 123
soybean 77, 137, 139, 193, 197
spine 180
squash 193
SRFA 132
SSR 114
stomata 35
strawberry 120
Streptomyces 46
STS 114
sulfonylurea 86
sunflower 43, 141, 143
sweet potato 122
systemic acquired resistance 25

T

T-DNA 20
tapetum 81
taxol 97, 106
Taxus 106
TEKTRAN 4, 9
TELNET 6
TEM 177
TEV 53
thaumatin 28
Theobroma cacao 119
thermocyler 145
Thielaviopsis basicola 42, 46
thymidine kinase 77, 81
Tifdwarf 141
Tifgreen 141
Tifway 140
Timopheevii 154
TMV 28
TNV 27
tobacco 27, 28, 36, 69, 72, 82, 86, 174, 198
tomato 13, 21, 31, 75, 81, 123, 142, 192, 197
toxin 17
transformation 14, 69, 72, 78, 119
transgenic 31, 60, 62, 79, 81
transient expression 82
transmembrane 168
transporter 169
Trichosanthes kirilowii 86

α-trichosanthin 86
tRNA consensus 158
TTIC 5
turfgrass 131
turnip 33
TVMV 53

U
untranslatable RNA 57
uptake 1 73
UV 31

V
vanilla 101
vegetables 112
Verticillium 31
vicilin 82
virus resistance 62
Vitis vinifera 120

W
watermelon mosaic virus 193
wheat 149
white pine 143
WMV II 54
world-wide-web 5
WWW/gopher 5

X,Y,Z
Xanthomonas campestris 29
yarn 185
yeast 73, 168
YOYO-1 179
Zoysiagrass 143
zucchini mosaic virus 193
ZYMV 53

Milton Keynes UK
Ingram Content Group UK Ltd.
UKHW051951071024
449327UK00026B/2267